AF679697

Systems Dependability Assessment

Systems Dependability Assessment Set

coordinated by
Jean-François Aubry

Systems Dependability Assessment

Benefits of Petri Net Models

Jean-François Aubry
Nicolae Brinzei
Mohammed-Habib Mazouni

ISTE

WILEY

First published 2016 in Great Britain and the United States by ISTE Ltd and John Wiley & Sons, Inc.

Apart from any fair dealing for the purposes of research or private study, or criticism or review, as permitted under the Copyright, Designs and Patents Act 1988, this publication may only be reproduced, stored or transmitted, in any form or by any means, with the prior permission in writing of the publishers, or in the case of reprographic reproduction in accordance with the terms and licenses issued by the CLA. Enquiries concerning reproduction outside these terms should be sent to the publishers at the undermentioned address:

ISTE Ltd
27-37 St George's Road
London SW19 4EU
UK

www.iste.co.uk

John Wiley & Sons, Inc.
111 River Street
Hoboken, NJ 07030
USA

www.wiley.com

© ISTE Ltd 2016
The rights of Jean-François Aubry, Nicolae Brinzei and Mohammed-Habib Mazouni to be identified as the authors of this work have been asserted by them in accordance with the Copyright, Designs and Patents Act 1988.

Library of Congress Control Number: 2015960014

British Library Cataloguing-in-Publication Data
A CIP record for this book is available from the British Library
ISBN 978-1-84821-991-5

Contents

Introduction

In the first book of this series [AUB 15], finite state automata were introduced as an efficient model for the study of reliability and dependability of systems as well in static as in dynamic context. We saw that this type of model requires either an *a priori* exhaustive knowledge of the possible states of the system or its formal construction by operations starting from the models of its components. This is unfortunately sometimes not possible. For example, during the design of a system these states are not known in advance. It is however useful to make a predictive dependability assessment in order to select the best solution among some propositions. Petri nets may be an interesting way to answer such problems. Widespread in the field of automatic control, especially for the modeling of discrete event systems, Petri nets were introduced in the field of dependability studies with a noticeable success. The objective of this book is not to present all of the forms of Petri nets used in dependability assessment but instead to focus on the most interesting ones. Before their description, we give a preliminary formal description of the different successive models of Petri nets which led to the advent of their use in the dependability field. Of course, it is not just a matter of exhaustively describing the existing variants of the basic models which are today hardly countable. In the same way, we will not demonstrate all the mathematical properties of these models and we will refer the reader to the essential basic works on the subject. After the introduction of the basic models called "autonomous Petri nets" and the comparison with the finite state automata especially in terms of event language expression, we will present the fundamental models of non-autonomous Petri nets to take account of the time and of an external environment, such models giving an opening to the study of hybrid systems. Relying on these timed and synchronized Petri nets, we will describe a systematic method of risk analysis based on an ontological approach whose elements are entities (supplier or target of hazard), their successive states and the events corresponding to these state changes. From the proposed model, a risk assessment may be deduced by simulation thanks to the introduction of random event generators. This approach is illustrated by an example from the railway transportation field. The need of models, integrating the stochastic character of elements (in this case, events) and allowing an analytical solution instead of simulation, leads to the introduction of stochastic Petri nets modeling and its equivalence conditions with Markov or some extensions of Markov models. We then show how, under some conditions, complex models may be simplified by a distribution of the global model on the two formalisms: stochastic Petri nets and Markov processes. Numerous extensions of Petri nets have been proposed; we recall the most significant ones and the conditions of their Markov process equivalence. To complete the book, we present some modeling examples using different available software tools. These examples are issued from different application domains.

Writing this book would not have been possible without the contribution of colleagues and of PhD and Master students who investigated some related aspects. All of these contributions have been the subject of publications and are referenced in the text. We would like to extend our thanks to G. Babykina, P. Barger, G. Deleuze, L. Gérard, R. Ghostine, D. Jampi, J. Lalouette, R. Schoenig, J-M. Thiriet and N. Villaume.

Jean-François AUBRY
Nicolae BRINZEI
Mohammed-Habib MAZOUNI
December 2015

Introduction to Part 1

Petri nets (denoted as PN in this book) were introduced by Carl Adam Petri in 1962 [PET 62]. As finite state automata (FSA) described in Volume 1 of this book series [AUB 15], PNs are intended to describe discrete event systems but contrary to FSAs, the transition function is explicitly described in PNs. Adding the suggestive and intuitive graphic representation, we can say that PN is a more powerful model than FSA to describe discrete event systems, due to the fact that an FSA may always be transposed into PN whereas PNs, for example, do not always have a finite state number. We will show here that the notion of language, set of all the possible event sequences in a system, may also be associated with a PN and that the class of these languages is wider than regular languages associated with FSAs.

Like for FSAs, PNs were the subject of multiple extensions at first to move them from the abstraction level, where only event sequencing is considered, to the level taking time into account. Timed PNs were defined to describe behavior of deterministic time systems. Following extensions, called non-autonomous PNs, associated with a PN, an external environment is needed in order to consider synchronization events, continuous variables, especially to describe controlled systems. All these models at various levels have an interest to model problems in the dependability assessment of systems.

1

Autonomous Petri Nets

1.1. Unmarked Petri nets

1.1.1. *Definitions*

A unmarked PN is a bipartite oriented 1-graph[1] provided with a mapping ϖ from the set of arcs to the positive integer set $\mathbb{N}^+$:

$$Q = \langle P, T, A, \alpha, \beta, \varpi \rangle$$

– P and T are two disjointed subsets of nodes: $P \cap T = \varnothing$:
  - P is the Place subset with a finite cardinal p;
  - T is the Transition subset with a finite cardinal t.

– A is the set of Arcs, α and β are the mappings associating with each arc, its origin and its goal nodes, respectively, so that:

$\forall a \in A, \quad$ if $\alpha(a) \in T$ then $\beta(a) \in P$
if $\alpha(a) \in P$ then $\beta(a) \in T$

– ϖ is a mapping or weighting function associating an integer with each arc, $\varpi : A \rightarrow \mathbb{N}^+$.

If $\mathbb{N}$ is reduced to $\{1\}$, the PN is of ordinary type (or state transition graph), otherwise the PN is of generalized type.

1 An oriented 1-graph is such that only one arc may be associated with a couple {origin node and destination node}.

Practically, rather than this formalism directly issued from the graph theory, we will use a definition where A does not explicitly appear. As it is an 1-graph, it consists of considering all the couples (P_i, T_j) or (T_i, P_j) and two applications w^- and w^+.

The PN is then defined as:

DEFINITION 1.1.– An unmarked PN or Place/Transition (P/T) net is a 4-uple $Q = \langle P, T, w^-, w^+ \rangle$ [DAV 89] where:

– P is the set of places (finite cardinal p);

– T is the set of transitions (finite cardinal t);

– $w^-(P_i, T_j) : P \times T \rightarrow \mathbb{N}$ is the backward transition function;

– $w^+(P_i, T_j) : T \times P \rightarrow \mathbb{N}$ is the forward transition function.

The value "0" associated with the couple (P_i, T_j) by w^- or w^+ means that there is no arc between P_i and T_j or T_j and P_i. If the value $k \in \mathbb{N}^+$ is associated with this couple by w^-, respectively w^+, then one arc oriented from P_i to T_j, respectively from T_j to P_i, exists between these nodes with the valuation k.

REMARK 1.1.– Another possibility is to define a PN as an n-graph (n arcs may exist between two nodes), an arc of weight n being replaced by n arcs each of them having the weight one.

1.1.2. *Drawing*

In the drawing of a PN, places and transitions are, respectively, represented by circles and streaks (or filled or empty rectangles) and the arcs are arrows to which the weights are attached. Figure 1.1 shows an example of PN with three places and two transitions respectively named as P_1, P_2, P_3, T_1, T_2. From this figure, we can write $w^-(P_1, T_1) = 1$, $w^-(P_3, T_2) = 2$, $w^-(P_2, T_1) = 2$, $w^+(P_3, T_1) = 2$, $w^+(P_1, T_1) = 1$, $w^+(P_2, T_2) = 3$.

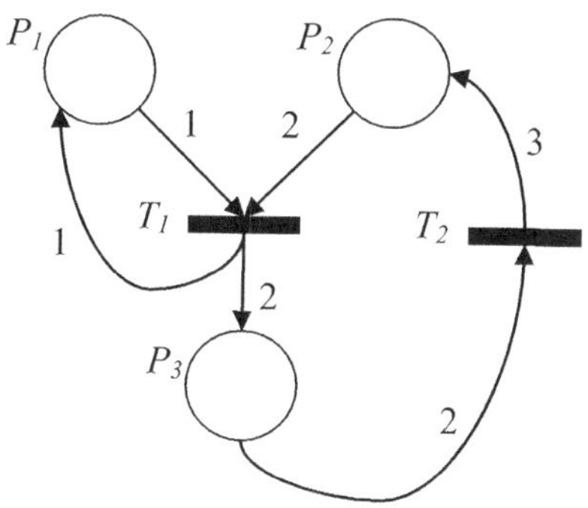

Figure 1.1. *The drawing of a PN*

1.1.3. *Other definitions*

Some of other definitions concerning particular cases of PN are summarized in the Appendix, section A.1.

1.2. Marking of a PN

Notations [DAV 89]:

– $I(T_j) = \{P_i \in P | w^-(P_i, T_j) > 0\}$ is the set of the input places of T_j;

– $O(T_j) = \{P_i \in P | w^+(P_i, T_j) > 0\}$ is the set of the output places of T_j;

– $I(P_i) = \{T_j \in T | w^+(P_i, T_j) > 0\}$ is the set of the input transitions of P_i;

– $O(P_i) = \{T_j \in T | w^-(P_i, T_j) > 0\}$ is the set of the output transitions of P_i.

For example, in Figure 1.1, $I(T_1) = \{P_1, P_2\}$ and $O(T_1) = \{P_3, P_1\}$.

The marking is a notion resulting from the association of tokens with the places of the PN. The position in the places of these tokens will evolve to represent the dynamics of the described system. This evolution is performed according to a set of rules described in section 1.3.

DEFINITION 1.2.– A marked PN is a couple $R = \langle Q, M_0 \rangle$ where Q is an unmarked PN and M_0 is an initial marking.

The marking M of a PN at a given instant is a p-sized columnar vector of integers (p is the place number of the PN), each of its component being the marking (or charge) of the place P_i that is to say the number of tokens inside P_i at the considered time instant:

$$M \in \mathbb{N}^p \qquad M^T = [M(P_1), M(P_2), \ldots, M(P_p)]$$

The initial marking M_0 is the marking at time $t = 0$.

Figure 1.2 shows the initial marking of the PN of Figure 1.1 with: $M_0^T = [1, 2, 0]$.

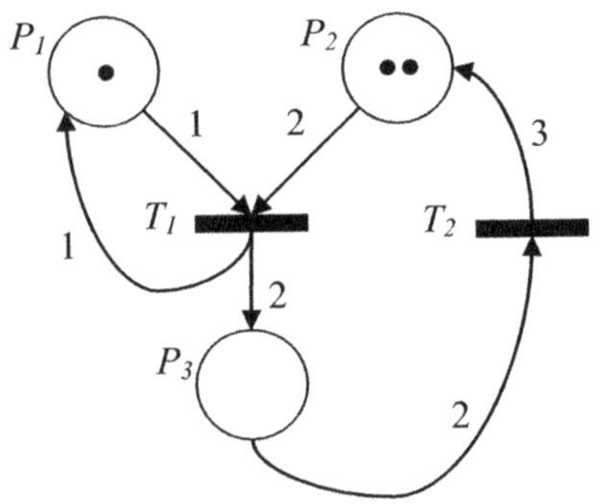

Figure 1.2. *A marked PN*

1.2.1. *Order relation on markings*

Let us consider two markings M_1 and M_2 of a PN.

We define the order relation between these markings as follows:

– $M_1 \geq M_2 \iff M_1(P_i) \geq M_2(P_i), \forall Pi \in P$;

– $M_1 > M_2 \iff M_1(P_i) \geq M_2(P_i), \forall P_i \in P$ and $\exists Pi | M_1(P_i) > M_2(P_i)$.

1.2.2. *Enabled transition*

The transition T_j is enabled for a given marking M if and only if:

$$M(P_i) \geq w^-(P_i, T_j), \forall Pi \in I(T_j)$$

In Figure 1.1, only the transition T_1 is enabled.

1.3. Dynamics of autonomous PNs

The previously defined notion of marking is the observation means of the evolution of the model. The position of the tokens will evolve according to a set of formal rules allowing the definition of some properties of the model. This will be recalled in the following, and more details may be found, for example, in [CAS 08, DAV 92, BES 01].

1.3.1. *Firing of a transition*

As PNs are models dedicated to discrete events systems, the firing of a transition may be considered as an event describing an elementary evolution of a system (see section 2.1 for the formal definition of labeled PN) characterized by the successive values of the marking before and after the firing. An enabled transition may be fired; from a given marking, each enabled transition could be fired but only one will be. The choice of the transition to be fired can be done arbitrarily. When a place has two output transitions their firings are in conflict. This notion of conflict (formally defined in Appendix A1.1) will be retrieved, for example, each time a failure occurs concurrently with a task activation or achievement. Some PN-dedicated software tools give the possibility of priority assignment to a transition concerned by a conflict, but this must be carefully handled to avoid the appearance of dead branches in the reachability graph. Two transitions $T_1, T_2 \in O(P_i)$ are not in conflict if they are not simultaneously enabled, which implies that these transitions have input places other than P_i.

The set of the enabled transitions must always be considered according to the current marking of the PN and not limited to a given place.

If M_b is the marking before the firing of T_j, the marking M_a after the firing is defined by:

– $\forall P_i \notin I(T_j) \cup O(T_j) \Longrightarrow M_a(P_i) = M_b(P_i)$

– $\forall P_i \in I(T_j) - (I(T_j) \cap O(T_j)) \Longrightarrow M_a(P_i) = M_b(P_i) - w^-(P_i, T_j)$

– $\forall P_i \in O(T_j) - (I(T_j) \cap O(T_j)) \Longrightarrow M_a(P_i) = M_b(P_i) + w^+(P_i, T_j)$

– $\forall P_i \in I(T_j) \cap O(T_j) \Longrightarrow M_a(P_i) = M_b(P_i) - w^-(P_i, T_j) + w^+(P_i, T_j)$

The firing of the transition T_j subtracts in place P_i as many tokens as indicated by $w^-(P_i, T_j)$ and adds in place P_k as many tokens as indicated by $w^+(P_k, T_j)$.

Figure 1.3 shows the PN of Figure 1.2 after the firing of transition T_2.

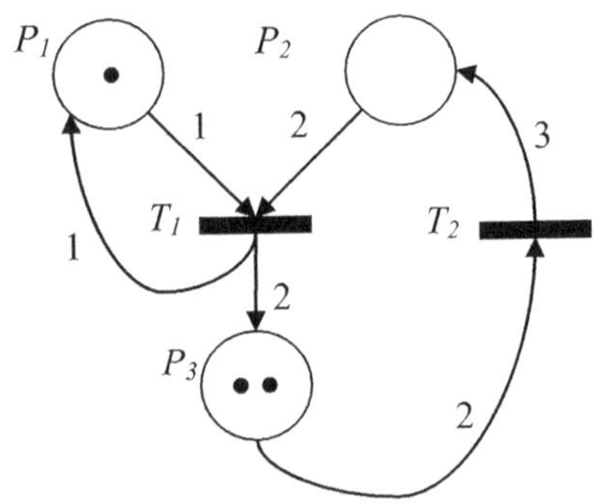

Figure 1.3. *PN of Figure 1.2 after firing of transition* T_2

1.3.2. *Transition matrix*

DEFINITION 1.3.– Let us define the backward matrix and forward matrix, as the following matrices with p lines and t columns:

$$W^- = \begin{bmatrix} w^-(P_1,T_1) & . & . & . & w^-(P_1,T_t) \\ . & . & . & . & . \\ . & . & w^-(P_i,T_j) & . & . \\ . & . & . & . & . \end{bmatrix} \quad [1.1]$$

$$W^+ = \begin{bmatrix} w^+(P_1,T_1) & . & . & . & w^+(P_1,T_t) \\ . & . & . & . & . \\ . & . & w^+(P_i,T_j) & . & . \\ . & . & . & . & . \\ w^+(P_p,T_1) & . & . & . & w^+(P_p,T_t) \end{bmatrix} \quad [1.2]$$

The transition matrix W is defined by:

$$W = W^+ - W^- \quad [1.3]$$

The transition matrix (p lines and t columns) is independent of the marking, each column simply shows the number of tokens to remove or add in a place when the corresponding transition fires. For Figure 1.1:

$$W = \begin{bmatrix} 1 & 0 \\ 0 & 3 \\ 2 & 0 \end{bmatrix} - \begin{bmatrix} 1 & 0 \\ 2 & 0 \\ 0 & 2 \end{bmatrix} = \begin{bmatrix} 0 & 0 \\ -2 & 3 \\ 2 & -2 \end{bmatrix}$$

1.3.3. *Firing sequence*

DEFINITION 1.4.– A firing sequence is obtained when a set of transitions are successively fired, starting from an initial marking. It is represented by the concatenation of the successive names of the fired transitions.

If for example starting from the initial marking M_0 the transitions T_1 then T_2 are fired to give the marking M_2, the sequence will be denoted as:

$$S = T_1 T_2, \qquad M_0 \xrightarrow{S} M_2$$

REMARK 1.2.– The transition set T provided with the concatenation operation and a neutral element may be considered as a monoïd denoted by T^*. With such a notation, a firing sequence is one element of this monoïd: $S \in T^*$. This notation will sometimes be used later.

DEFINITION 1.5.– Let S be a firing sequence feasible from a marking M_i, the characteristic vector of the sequence denoted as N is a t size vector, whose j^{th} component represents how many times the transition T_j is fired in the sequence S:

$$N \in \mathbb{N}^t, \qquad N(T_j) = k \text{ if } T_j \text{ is fired } k \text{ times.}$$

1.3.4. *Reachable marking*

The M vector cannot take any value. From a given marking M_0, it is possible to list all the possible firing sequences. The obtained marking after each of these sequences is a reachable marking.

Let us note that $R(M_0)$ is the set of the reachable marking from the initial marking M_0:

$$R(M_0) = \{M_a \in \mathbb{N}^p : \exists S / M_0 \xrightarrow{S} M_a\}$$

1.3.5. *Fundamental equation*

In FSA, we defined the state changes by the mean of the transition function. In PNs, this function is defined as follows:

$$f : \mathbb{N}^p \times T \to \mathbb{N}^p$$

$f(M_k, T_j)$ is defined if and only if T_j is enabled, in this case, $f(M_k, T_j) = M_{k+1}$ with:

$M_{k+1}(P_i) = M_k(P_i) - w^-(P_i, T_j) + w^+(T_j P_i)$ for $P_i \in I(T_j) \cup O(T_j)$

As for the FSAs, we can extend f from the domain $\mathbb{N}^p \times T$ to the domain $\mathbb{N}^p \times T^*$ (T^* being the monoïd on the set T provided with the concatenation operation (see section 1.3.3)) and define for a given initial marking, the new obtained marking after a firing sequence of characteristic vector N.

We then obtain the fundamental matrix equation as:

$$M_k = M_i + W \cdot N \qquad [1.4]$$

For Figure 1.2, let us imagine from the initial marking, the firing sequence T_2T_1. After the firing of T_2, the obtained marking is shown by Figure 1.3 and after the firing of T_1 it becomes as indicated by Figure 1.4.

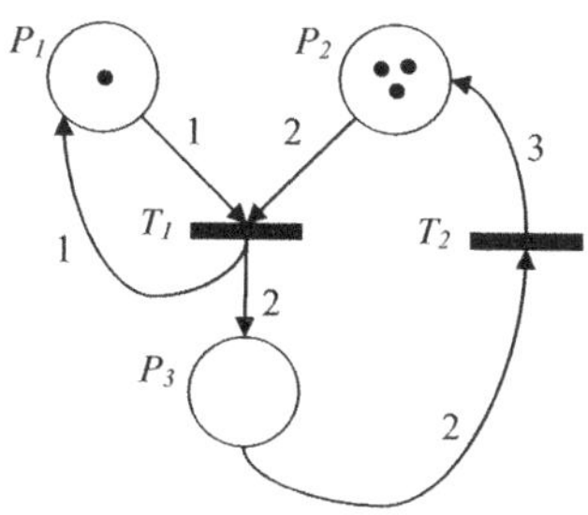

Figure 1.4. *PN state of Figure 1.3 after firing of transition T_1*

As the two components of the vector N are 1 and 1, each of the two transitions being fired one time, the obtained marking may be retrieved by the following calculus:

$$\begin{bmatrix} 1 \\ 3 \\ 0 \end{bmatrix} = \begin{bmatrix} 1 \\ 2 \\ 0 \end{bmatrix} + \begin{bmatrix} 0 & 0 \\ -2 & 3 \\ 2 & -2 \end{bmatrix} \cdot \begin{bmatrix} 1 \\ 1 \end{bmatrix}$$

1.3.6. *Properties of PN*

A set of definitions and properties are summarized here. For a complete description and formal demonstrations of properties, we can report to [DAV 89, CAS 08, DAV 92, BES 01, BRA 83]:

– *Boundedness:*

- a place of a PN is bounded for a given initial marking M_0 if for any accessible marking from M_0 the token number in this place remains finite. If $\forall M_n \in R(M_0)$, $M_n(P_i) \leq k$ with $k \in \mathbb{N}$, then P_i is k-bounded,

- a PN is bounded for a given initial marking M_0 if all the places are bounded for M_0.

If $\forall P_i \in P, \forall M_n \in R(M_0)$, $M_n(P_i) \leq k$ with $k \in \mathbb{N}$, then the PN is k-bounded.

These properties are dependent of the initial marking but sometimes a PN may be structurally bounded, that is to say bounded for any initial marking.

– *Liveness:*

- a transition T_j is alive for a given marking M_0 if $\forall M_n \in R(M_0), \exists S : M_0 \xrightarrow{S} M_n / T_j \in S$ (there is always a firing of T_j),

- a PN is alive for a given marking M_0 if all its transitions are alive for M_0.

– *Blocking:*

- a blocking is a marking from which any transition is enabled. It corresponds to an absorbing state,

- a PN is blocking free for a given initial marking M_0 if no marking $M_n \in R(M_0)$ is a blocking.

Liveness and blocking are properties dependant on the initial marking M_0.

1.3.7. *Other properties*

Some other properties are summarized in Appendix A.1.

1.3.8. *Invariants in a PN*

1.3.8.1. *Conservative component and marking invariant*

It should be noted that sometimes the weighted sum of the markings of a subset of places remains constant. This is an invariant of this subset which is called conservative component of the PN. As it is independent of the initial marking, this is a property of the unmarked PN (the value of this constant may only depend on the initial marking). In most cases, this is the characteristic of a physical property of the modeled system.

A P-semi-flow is a vector F of integers of dimension p (number of places of the PN) so that:

$$F^T \cdot W = 0 \qquad [1.5]$$

According to the fundamental equation $M_k = M_i + W \cdot N$ (for any accessible marking from M_i by a firing sequence S characterized by the vector N): $F^T \cdot M_k = F^T \cdot M_i + F^T \cdot W \cdot N$. If $F^T \cdot W = 0$ we obtain:

$$F^T \cdot M_k = F^T \cdot M_i \qquad [1.6]$$

which is the marking invariant.

The integers of the vector F may be considered as weights assigned to the places of the PN. The subset of places whose weights are null is the PN conservative component support of the P-semi-low. It will be noted P_F.

Any linear combination of a semi-flow is itself a semi-flow.

Let $P_F = \{P_1, P_2, ..., P_r\}$ be a conservative component of a PN and $F = [q_1, q_2, ..., q_r]^T$ the corresponding weighting vector. All the places of the conservative component are bounded and we get: $M(P_i) \leq F^T \cdot M_0/q_i$.

For example, it is easy to verify that in the PN of Figure 2.1 (see section 2.2) the subset of places $\{P_4, P_5\}$ is a conservative component, the sum of their marking is always equal to 1 (the initial marking of P_4).

1.3.8.2. *Repetitive component and firing invariant*

In the same way, a T-semi-flow is defined: $W \cdot F = 0$.

Here, the weighting vector of integers is a vector N (dimension = t, N_i being the firing number of T_i) associated with a transition sequence S. Let us note T_S the transition subset fired at least once in the sequence S.

T_S is a stationary repetitive component if and only if

$$W \cdot N = 0 \qquad [1.7]$$

T_S is an increasing repetitive component if and only if $W \cdot N > 0$.

If $W \cdot N = 0$, then N is a T-semi-flow but any semi-flow does not necessarily correspond to a repetitive component because it must correspond at least to a firing sequence.

If S is a repetitive sequence from the marking $M_1 \in R(M_0)$ and if S is also a firing sequence from $M_2 \in R(M_0)$, then S is also a repetitive sequence from M_2 (see fundamental equation).

For Figure 2.1 (see section 2.2), the transition set $\{T_2, T_3\}$ is a repetitive component because a firing of T_2 leads to a firing of T_3 and so on.

1.3.9. *Reachability graph*

The evolution of the marking due to transition firings may be represented by a graph called a reachability graph.

The reachability graph of a PN $RG(M_0)$ is a graph whose nodes are associated with the successive values of the marking vector from initial marking M_0 and whose arcs correspond to the firings of transitions. All the properties of a PN may be retrieved on the reachability graph.

Let us consider the PN of Figure 1.5 (left) and its initial marking with two tokens in place P_1 and one in place P_3. The corresponding reachability graph is on the right-hand side of the figure. The node $[2, 0, 1, 0]^T$ corresponds to the initial marking that may evolve by firing of transitions T_1 or T_2 to reach respectively the markings $[1, 1, 1, 0]^T$ and $[2, 0, 0, 1]^T$, and so on.

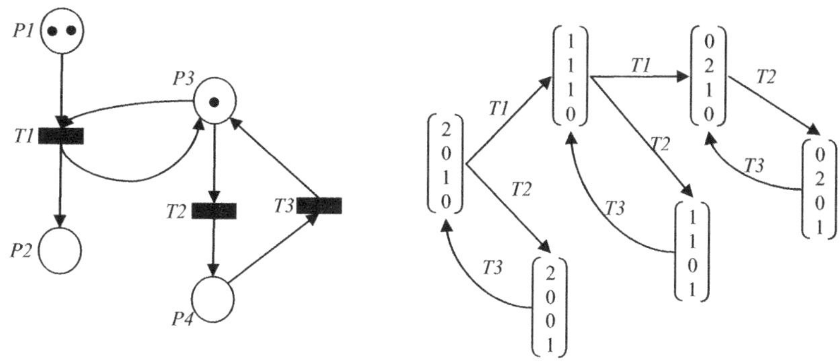

Figure 1.5. *A marked PN and its reachability graph*

In the current example, the reachability graph is finite but it is possible that it is not the case, meaning that the PN is not bounded (see section 1.3.6). It is then possible to define a finite covering graph by the identification of cycles in the reachability graph [DAV 89, DAV 92, DIA 01].

2

Petri Nets and Event Languages

The concept of event language, its representation by finite state automaton (FSA) and its ability to characterize and evaluate the dependability of a system was the subject of the first book [AUB 15] of this series. The problem is now to show how the PN model may be associated with the concept of event set characterizing a system and to compare the expression power of the obtained model to that of FSAs. To do that, let us introduce the concept of labeled Petri net.

2.1. Labeled PNs

As we associated a finite state automata (FSA) (a graph) with the event set E characterizing a given system, a labeled PN [CAS 99] is the association of a PN with this event set. To each transition of the PN, a mapping associates an event of E.

2.1.1. *Formal definition*

DEFINITION 2.1.– A labeled PN is a 4-tuple $H = \langle R, E, l, P_m \rangle$ where:

– R is an autonomous marked PN $\langle P, T, w^-, w^+, M_0 \rangle$;

– E is the event set (including the neutral event ε);

– l is a mapping from T to E, called labeling function of the transitions

– $P_m \subset P$ a subset of places called final places[1].

2.1.2. *Generated and marked languages*

The generated and marked languages of a labeled PN may then be defined as [CAS 08, CAS 99]:

– generated language: $L(H) := \{l(S) \in E^* | S \in T^* \text{ and } M_0 \xrightarrow{S} M_a$ is defined$\}$;

– marked language: $L_m(H) := \{l(S) \in L(N) | S \in T^* \text{ and } M_0 \xrightarrow{S} M_a$ is defined such that $\forall P_i \in Pm,\ P_i$ is marked in $M_a\}$;

where l is in its extended form: $l : T^* \rightarrow E^*$ (see section 1.3.3).

These definitions are coherent with the corresponding ones defined in [AUB 15] for the FSA.

The class of languages representable by a labeled PN is [CAS 99]:

$$\begin{aligned} L_{PN} : \ &= \ \{K \subseteq E^* | \exists H = \langle P, T, w^-, w^+, E, l, M_0, P_m \rangle, \\ &\quad [L_m(H) = K]\} \end{aligned}$$

The properties are of course dependant of l (whether or not it is injective) or of Pm (whether or not it is finite).

REMARK 2.1.– This model may be compared to the notion of synchronized PN (see section 4.3) introduced by Moalla [MOA 78b, MOA 78a]. The difference between these two models lies in the nature of events. In synchronized PNs, considered as non-autonomous PNs, the events are external entities (acting like "inputs") coming from the environment of the modeled discrete event system (DES). However, for labeled PNs, there is no particular

1 The qualifier "marked" commonly used in FSA must be avoided not to confuse with the marking of PNs.

hypothesis on the nature of events like in the FSA model (nevertheless, the events of labeled PNs are sometimes considered as "outputs" that is to say "produced" by the model).

In the reachability graph of a labeled PN, each arc corresponding to a transition firing is consequently associated with an event.

2.2. Example

Let us consider in a manufacturing system, an assembly operation on parts. There are two types of parts to be assembled and they are available in two input stocks. Two parts of type 1 and one part of type 2 are to be assembled together. After the operation, the obtained part is filed in an output stock. Let us suppose, in addition, that the assembly machine may be unavailable for the operation after its failure occurrence. Fortunately, a repair action may occur to make the machine available again. The PN of Figure 2.1 models these processes with 5 places and 3 transitions.

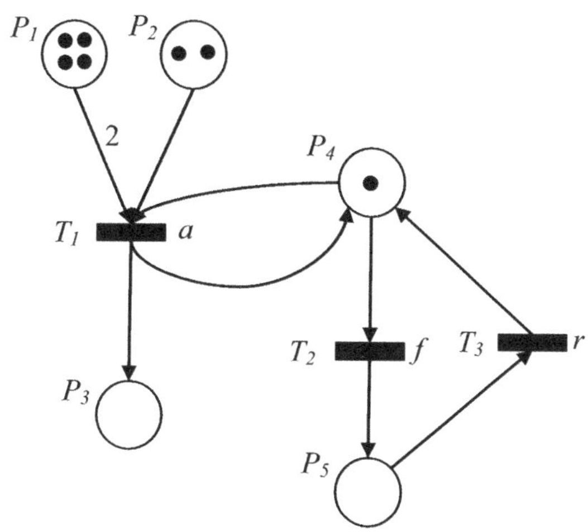

Figure 2.1. *PN of an assembly system*

P_1 and P_2 represent the stocks of parts to be operated and P_3 represents the stock of parts assembled. The number of tokens in these places equals the number of parts in the stocks. P_4 and P_5 are places representing, respectively, the availability and the unavailability of the machine. Transition T_1 describes the assembly operation, T_2 corresponds to the failure event of the machine and T_3 its repair event. Let us suppose that at the initial time, four parts of type 1 and two parts of type 2 have been filed in the input stocks, that the output stock is empty and that the machine is available. This means that in the PN model, places P_1, P_2 and P_3 are marked respectively with four, two and one token and the other places are unmarked. The initial marking is then $M_0^T = [4, 2, 0, 1, 0]$. From this initial marking, transition T_1 is enabled and may then be fired. This firing takes, respectively, two, one and one token in places P_1, P_2 and P_4 and sets down one token in place P_3. The new marking vector becomes: $M_1^T = [2, 1, 1, 1, 0]$. Transition T_1 is always enabled but transition T_2 too. If T_2 is fired in case of the machine failure, T_1 becomes enabled and a second assembly cannot be processed until the repair of the machine; the corresponding marking is $M_2^T = [2, 1, 1, 0, 1]$. As soon as the machine is repaired, a second assembly may be processed and the marking becomes: $M_3^T = [0, 0, 2, 1, 0]$ where P_1 and P_2 are unmarked. Transition T_1 becomes disabled and no other assembly can occur.

The generated language of this PN comprises all the possible transition sequences from the initial marking. If a, f and r are the events labeling the transitions, this language contains the sequences a, aa, $aa(fr)^*$ indicating that after two assemblies the machine can fail and be repaired infinitely but also contains all the sequences including alternations of fr before and after an occurrence of a: $L = \{(fr)^*, (fr)^*f, (fr)^*a(fr)^*, (fr)^*a(fr)^*f, (fr)^*a(fr)^*a(fr)^*, (fr)^*a(fr)^*a(fr)^*f\}$ (do not forget about the neutral event $\varepsilon \in (fr)^*$). Let us note that, as the definition suggests, the language is dependent of the initial marking. In our case, the initial marking limits the occurrence number of a.

Nevertheless, this simple example lacks realism or precision. It only describes the possible event sequences, the firing of a transition being instantaneous by definition. This model implicitly considers, for example, that machine failure cannot occur during the assembly process. The need for a representation of time appears to be essential. As failure and repair events are essentially of random type, the introduction of stochastic behavior in the model becomes essential too to access the quantitative assessment of the system performances. It may be considered in addition that the assembly operation of the example also has a random duration but in a time scale very small with regard to failure or repair of components in the system. This will not be a problem in the frame of stochastic PN which we will introduce later on. However, in the case of large systems, the presence of different time scales will be used to simplify the whole model. All these extensions of PNs will be considered in this book.

3

Comparison Petri Nets – Finite State Automaton

Both models are used to represent the evolution of the system states according to the event occurrences. In an FSA, the whole set of state must be first defined and the state changes (the arcs) defined by the transition function. For large systems, this may be a fastidious task but it is possible to build the whole model from elementary submodels by means of special operators (especially the synchronization operator). The building procedure of an FSA is more like an analysis process. The system must be perfectly known. In a PN, the transition function explicitly appears on the drawing of the model through the concept of transition enabled but the system states do not appear explicitly; they are distributed on the set of places by means of tokens. The building of a PN is more relevant to a conception process. The places may have different meanings, such as stock parts, activity or not of an entity of the system, shared resources, synchronization of actions, etc. The model is more close to the "physical" behavior of the system. There is no rule that says to use one model instead of the other; it is more a matter of purpose (type of problem), habit, constraint or a lack of knowledge on system state numbers. However, there are some important differences between these models and it is essential to understand these.

3.1. Language expression

In the first book of this series [AUB 15], we showed the ability of FSA to represent languages. However, this is limited only to regular ones[1]. For example, let us consider the language $L = \{\varepsilon, ab, aabb, aaabbb, \ldots\}$ which is also described as $L = \{a^n b^n : n \neq 0\}$, based on the alphabet $E = \{a, b\}$. This language may only be represented by the arborescent automaton of Figure 3.1. In this automaton, the horizontal path "memorizes" the number of successive recognized letters "a" and each slantwise path marks an identical number of letters "b". Considering that L may be limited to a finite value of n, the automaton is finite and a different automaton corresponds to each different value of n. If n becomes infinite, the automaton also becomes infinite.

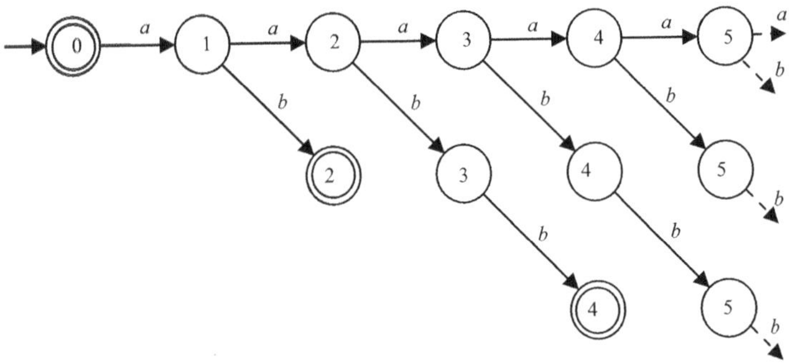

Figure 3.1. *Arborescent automaton equivalent to the language* $a^n b^n$

Figure 3.2 shows a labeled PN generating the language $\{a^n b^n\}$. Place P_3 memorizes the number n of successive firings of T_1 labeled by a. As soon as T_2 is fired once (memorized by P_2), T_1 cannot be fired and T_3 only will be fireable $n-1$ times. The PN is then blocked.

Whatever the value of n, the PN structure is always the same; only the marking differs. When n becomes infinite, the marking of place P_3,

1 The definition of a regular language call only on the operations union, concatenation and iteration.

particularly, becomes infinite too, indicating that the PN is unbounded. It is easy to verify that the automaton of Figure 2.1 is perfectly homogeneous to the marking graph of the PN of Figure 3.1.

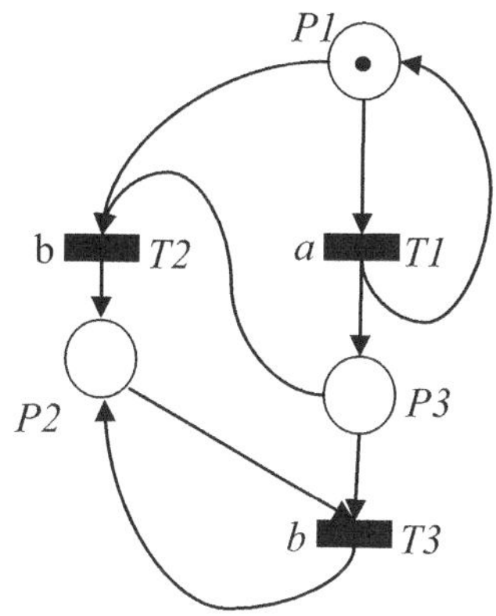

Figure 3.2. *Labeled PN equivalent to the language* $a^n b^n$

This example shows that PN formalism can characterize more languages in E^* than FSA formalism can. It is clear that this advantage comes from the ability of PNs to memorize actions by means of tokens that may be accumulated in a place. It has been demonstrated by [CAS 99] that the class of languages representable by PNs is strictly greater than those representable by FSAs in the case of models needing a finite number of memories. A finite state automaton may always be equivalent to a PN generating and marking the same languages, but the inverse is not true.

3.2. Building of the models

A system behavior may be described by several disconnected PNs but there is no equivalent of the synchronization operation (or parallel composition) defined on FSAs. The link between these PNs is made by the marking vector containing the whole set of places of all of them.

In case of labeled PNs, if the different sub-PNs do not share any event, the model remains constituted of these disconnected PNs. In case of event sharing, it will be necessary to build one by one the convenient mechanisms.

In the following sections, we will show it is easy to assemble two PNs together with some simple examples.

3.2.1. *Synchronization of submodels*

Defining a synchronization operator between two labeled PNs sharing a subset of events, as we did with FSA (see [AUB 15]), is not possible. Considering the reachability graphs of both PNs, we might think that it would be sufficient to reverse the operator defined on FSA. However this is not possible for at least two reasons, the first being the possible non-finiteness of the reachability graph and the second being its dependency on the initial marking.

Nevertheless, many solutions for PN synchronization have been proposed, albeit outside the context of labeled PNs. We will present some mechanisms often used, for example, in the modeling of information systems with autonomous PNs. Let us consider a given system being modeled by two disconnected autonomous PNs (two tasks or functions concurrently executed). In the first PN, let us consider the transition T_1 between places P_1 and P_2, and in the second PN, the transition T_2, between places P_3 and P_4. Let us suppose now that the two tasks have to be executed such that transition T_2 can not be fired before T_1. This condition can be achieved as shown in Figure 3.3(a) by the introduction between T_1 and T_2 of the intermediary place P_5 which will be marked on the firing of T_1 allowing then the firing of T_2. This mechanism is sometimes called "semaphore".

A second example concerns the case where the two transitions T_1 and T_2 must be simultaneous fired. It is the concept of "rendez-vous". The PN of Figure 3.3(b) shows that both transitions can be merged in one sole T' that will be fired when both places P_1 and P_3 are marked.

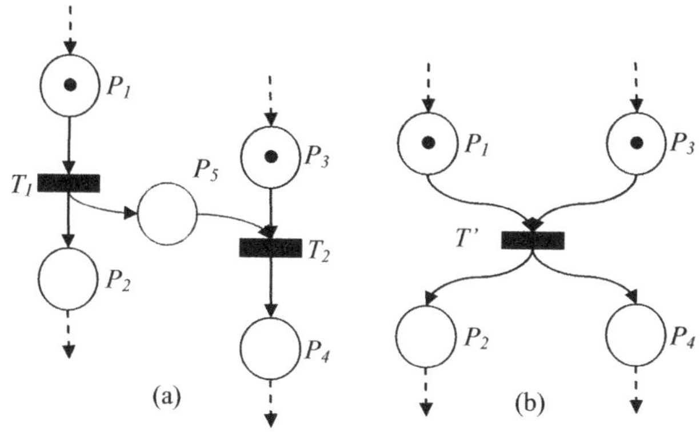

Figure 3.3. *Simple case of two PNs synchronization*

Despite these two simple examples, the synchronization of two autonomous PNs must be examined case by case and any assembly of several PNs will be followed by a checking of the properties (liveness, reachability, etc.) of the obtained model.

Practically, modeling the behaviour of a physical system requires non-autonomous PN models that take the "environment" into consideration. This environment may include events, variables, functions and predicates, any of which can be used to implement a synchronization mechanism. Some of these models will be summarized later in sections 4.3, 4.5 and 12.2.

3.2.2. *Resource sharing*

When two disconnected PNs describing two subsystems contain a place modeling the same entity of the system, these places can be merged to create a common place. This is the case, for example, for the problem of the resource sharing by two (or more) processes. For example, two machines, two computers or two workers share a common resource, an operator, a memory or a tool, which is necessary to perform their operation. Let us take the case of two machines sharing the same operator for a repair action when they have failed.

In Figure 3.4, P_1, P_3 and P_5, when marked, indicate that both machines are operating and that the repair operator is available. Both transitions T_1 and T_4 are enabled, and one of them will be fired, for example T_1 on the failure of the first machine. P_2 becomes marked and T_2 becomes enabled. The repair of the machine can be executed as the repairer is available (T_2 enabled). However, if the second machine is being repaired, the repair of the first machine must wait until the release of the repair operator. The problem can of course be extended to the case of other machines sharing the same resource.

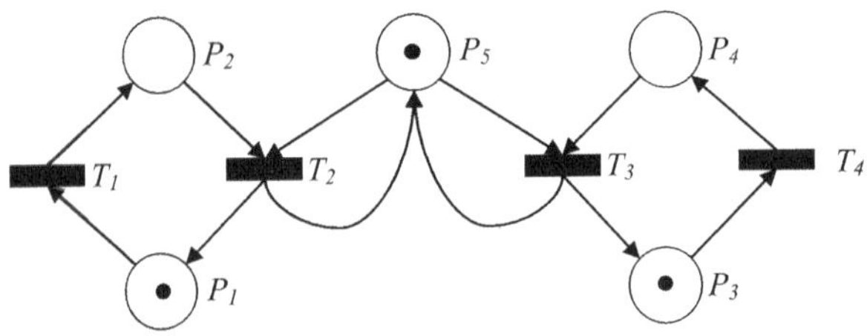

Figure 3.4. *Resource sharing between two sub-PNs*

Let us remark that the model does not give information about a potential priority to the repair of one machine rather than to the other one. To find out, it is necessary to introduce additional places and transitions to distinguish the beginning and end events of the repair action materialized by a specific place whose token may be retrieved by a priority request. One of the main advantages of PNs is being able to model the need of memories by means of tokens in places. Let us recall that with FSA this is only possible by adding as many states as necessary. For example in the dependability field, a Markov graph may be transformed by the fictive state method to include the need of a memory.

3.2.3. *Construction by refinement*

Unlike with the FSAs, there is no formal operation between PNs and it is only possible to interconnect them by means of mechanisms such as those presented above. It is then recommended to verify the properties of the results because the reiteration of such mechanisms without precaution can lead to non-liveness, non-bounding, blocking, etc.

An interesting approach consists in building PN by means of refinement primitives allowing to obtain *a priori* good properties, particularly in liveness, as a result. This approach, introduced in [TAN 88] and developed in [MOI 91] or [AUB 91], has been used in the automotive field for the dependability assessment of automated subassemblies [JAM 01]. The concerned PNs are "control PNs" where a place is associated with a control task to be executed. Practically, these PNs are state graphs (unique token) and each PN is a description of a sequence control including reaction to failures of sensors and actuators. The whole control system is described by disconnected PNs potentially synchronized.

Three refinement primitives [AUB 08] are proposed (Figure 3.5).

The embryo primitive (EMB) is an embryonary PN used to define the initial specification of an iterative process which periodically performs a global operation modeled by a single place. This operation will then be transformed in increasingly simple operations until obtaining elementary ones. The sequence primitive (SEQ) allows the decomposition of an operation into two suboperations always executed in the same order. To model this, one place and one transition are introduced in the PN. The fork primitive (FOR) gives the possibility to link two places of the same PN by a transition. It allows for the introduction of alternatives as well as upstream or downstream connections in the PN.

The construction is established step by step with the primitives as follows:

– the (EMB) primitive is used to initialize the construction and then this primitive will no longer be used;

– the PN is developed using the primitive (SEQ) and then (SEQ) again or (FOR) until the whole description is obtained.

Figure 3.6 illustrates this approach using an abstract example.

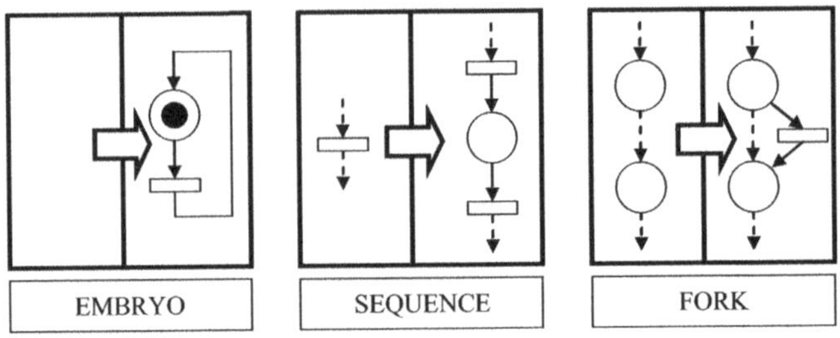

Figure 3.5. *The three construction primitives*

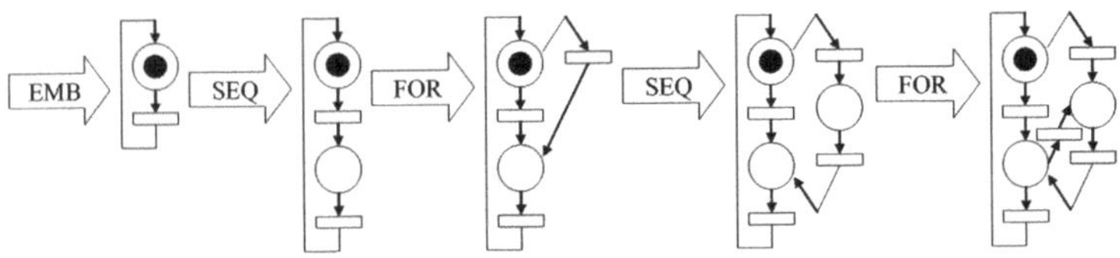

Figure 3.6. *Application example of the primitives*

This construction method of PNs by the use of primitives has the disadvantage of making modifications difficult. Indeed, a modification in a part of the PN obtained at the step n implicates the resumption of the construction at the step $n - 1$ and the redevelopment of all the following steps. This implicates the use of support tools keeping in mind all the successive steps of the development. Nevertheless, this drawback is balanced by the properties of the resulting PN (especially liveness).

3.3. Compactness of the model

The compactness of a model is a criterion to define the memory size necessary for this model to be stored in computer systems. Both models FSA and PN are graphs and can therefore be easily compared, the size of the requisite memory to store such models being directly linked to the number of elements (nodes and arcs). An advantage is then given to PN due to the marking mechanism. If now we talk about simulation, the compactness of the model has no real impact on the simulation time as it is evident that for any of the two models, all the system states must be explored and this is not dependent of the model type but only of the modeled system.

Until now we have only considered the problem of characterizing the possible event sequences in a system without making considerations on the origin of these events or on the time between them. These considerations have been the subject of a lot of extensions of the basic PN model as we will see now in the following pages.

4

Some Extensions of Petri Nets

Some Petri net (PN) extensions among the set of all those exist will now be presented. We made the selection with the care of showing their interest in dependability purposes: either they are directly used by the experts or they constituted intermediary steps toward actual standards such as stochastic PNs.

4.1. PN with inhibitor arcs

Despite the loss of structural properties, it is useful to describe this type of PN because it is sometimes proposed in various software tools dedicated to dependability studies.

To an autonomous PN, a new type of place–transition arc $P \to T$ is added. An inhibitor arc, such as the arc between P_i and T_j in Figure 4.1, signifies that the transition T_j can only be fired if the place P_i is unmarked. No w transition function is associated with this arc; for the firing of the transition T_j, the place P_i does not contribute to the marking evolution. While this property may be interesting for modeling some mechanisms, it can in contrast induce the loss of the structural properties of the PN (it is sometimes possible to translate a PN with inhibitor arcs to an ordinary PN).

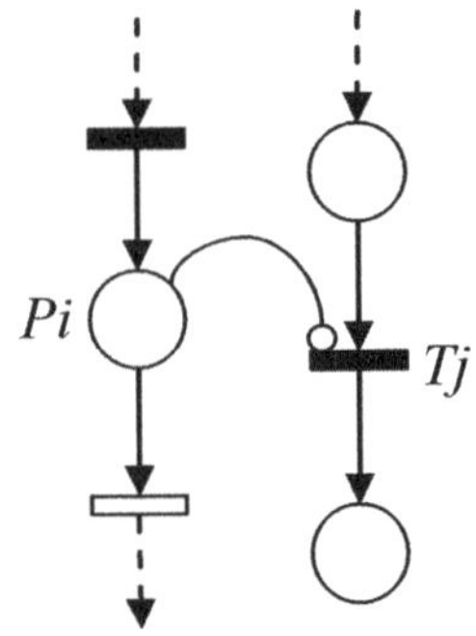

Figure 4.1. *PN with inhibitor arc*

4.2. Timed PN

Timed PNs are not expressly used in dependability modeling, however, a short presentation is useful because they can be considered as a transitory model for delivering the stochastic one that will be presented in Chapter 9. If we refer to Figure 3.4, the presented model did not describe the time duration of the repair process, that is to say the unavailability of the repairer for the eventual repair of the other machine. Transition T_2 representing the repair action in this example must then be refined and replaced by a first transition for the repair start, a place representing the duration of the repair and and a last transition to represent the repair end. To the additional place, a predefined sojourn time for the token may be associated with this place. In the same way, it is foreseeable to introduce a duration to firing of a transition.

Timed PNs (denoted by TPN) have been introduced to take account of the time by its association to the nodes of the graph to represent the time sojourn in a place or the firing time duration of a transition.

In the P-timed PN introduced by Sifakis [SIF 77, SIF 80], time delays are associated with places and concern the availability of the token, when it enters a place; it will be only available after this time delay. As soon as the delay is elapsed, the token becomes available to enable the downstream transitions with the same rules as for autonomous PNs.

In the T-timed PN introduced by Ramchandani [RAM 74], time delays are associated with transitions. When a token enters a place, it immediately contributes to the transition firing. However, the token will be set down in the place of arrival only after a given time delay. The transition is then fired in two times. It has been demonstrated [DAV 92, CHR 83] that the translation of one of these models to the other one is possible meaning that they are equivalent.

4.2.1. *P-timed Petri nets*

DEFINITION 4.1.– A timed Petri net is a couple $R_T = \langle R, \tau \rangle$ where:

– $R = \langle P, T, w^-, w^+, M_0 \rangle$ is an autonomous PN;

– τ is a mapping from P to $\mathbb{Q} \cup \{0\}$ such that: $\forall P_i \in P, \tau(P_i)$ is the delay associated with the place P_i, $\mathbb{Q}$ being the set of rational numbers.

EVOLUTION 4.1.– When a token enters a place P_i, it remains unavailable for a time $\tau(P_i)$ and cannot be considered to enable transitions. When the time $\tau(P_i)$ is elapsed, the token becomes available and can contribute to enable the downstream transitions. The firing process is then the same as for autonomous PNs.

REMARK 4.1.– The marking of each place can be divided into two parts, one defining the number of unavailable tokens (M_u) and the other defining the number of available tokens (M_a).

4.2.2. *T-timed Petri nets*

DEFINITION 4.2.– A T-timed Petri net is a couple $R_T = (R, \delta)$ where:

– $R = \langle P, T, w^-, w^+, M_0 \rangle$ is an autonomous PN;

– δ is a mapping from T to $\mathbb{Q} \cup \{0\}$ such that $\forall T_i \in T, \delta(T_i)$ is the delay associated with the transition T_i, $\mathbb{Q}$ being the set of rational number (≥ 0).

This model is used for the representation of system states whose changes correspond to operations of non-zero duration as, for example machining or data processing operations.

EVOLUTION 4.2.– A token may or may not be reserved for the firing of a transition. When a transition T_j is fired, non-reserved tokens are set down in the places $O(T_j)$ according to the weighting functions of the concerned arcs. Then, they contribute to enabling the downstream transitions of $O(T_j)$. The transitions to be fired are then selected and the tokens enabling them are reserved. The effective firing of a fireable transition T_i will occur $\delta(T_i)$ later unless the firing of another transition before this time disables T_i . At the initial time, any token is reserved.

4.3. Synchronized PN

We already mention this type of PNs in relation with labeled ones. They were introduced by Moalla *et al.* [MOA 78b, MOA 78a].

This model is considered as a non-autonomous PN that is to say a PN whose evolution is not only defined by the marking but also dependent on an external environment composed here by a set of events acting as "inputs" on the system. Let us recall the formal definition:

DEFINITION 4.3.– A synchronized PN is a 3-tuple $Rs = \langle R, E, \mu\rangle$ where:

– R is a marked PN $R = \langle P, T, w^-, w^+, M_0\rangle$;

– E is a set of external events;

– μ is a mapping from T to $E \cup \{\varepsilon\}$, ε being the event always occurring (neutral element of the monoïd E^*).

On the graph, $\langle\mu(T_j)\rangle$ is written close to the transition T_j to specify the event associated with T_j.

The transition T_j is receptive to the event e for a given marking M_i, if and only if T_j is enabled by M_i and $\mu(T_j) = e$.

When enabled, a transition will be fireable when the associated event will occur, then it is immediately fired (except eventually in case of conflict). If the event associated with a transition is ε, the transition is fireable as soon as it is enabled.

The notion of reachable marking is distinguished from the notion of stable marking waiting for a synchronization event $\in E$.

A marking M_i of Rs is said to be stable if no transition of Rs is receptive to $\{\varepsilon\}$ for M_i.

When the event e_k occurs, the transition receptive to it is fired and then all the transitions enabled by this firing and receptive to ε are fired too until a stable marking be obtained. This evolution on e_k occurrence is called iterated firing sequence on e_k occurrence.

This synchronization of the PN can modify the properties inherent to the markings of the underlying autonomous PN.

When used as a simulation support, a synchronized PN must be completed with the environment model that may be achieved by another PN whose transition firings are considered as event emission. If the software tool used for the simulation does not include the concept of synchronization by events, it may be replaced by a simple mechanism derived from Figure 3.3(a). Although synchronized PNs do not explicitly refer to time, this one appears through the event concept. An event is defined by its attributes such as its nature, occurrence date and occurrence number so that for any event emission by an entity its consumption by another entity must be precisely described. In our case, the mechanism must be such that the firing of an enabled transition in the synchronized PN must be simultaneous with the firing of the emitter transition. In Figure 4.2, transition T_1 is the event emitter in the first PN and transition T_2 is the transition to be synchronized in the second PN. The place P_5 and the transition T_3 are inserted between the two initial PNs to implement the mechanism. On the firing of T_1, a token is deposited in P_5. If P_3 is marked, transition T_2 is instantaneously fired but if it is not the case, P_5 remains marked as long as P_3 is not. T_2 can then be fired at an undefined time after T_1. To avoid this, a "well" transition T_3 is inserted as output of P_5 to consume the token indicating that the synchronization failed. When P_3 is marked again, the firing of T_2 will wait for a new occurrence of the event. When it comes, P_5 is marked again, and with T_3 and T_2 being in conflict, it is necessary to assign a priority of T_2 over T_3.

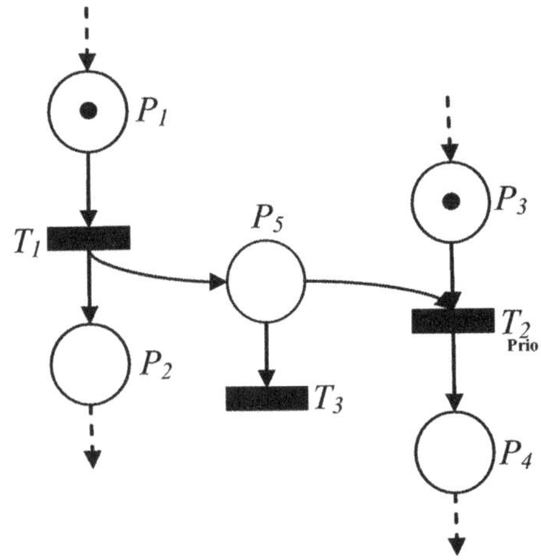

Figure 4.2. *Synchronization mechanism*

4.4. Timed synchronized PN

These PNs are an evolution of synchronized PNs where the time is introduced to characterize the sojourn of tokens in the places. They were introduced by Moalla *et al.* [MOA 80, MOA 85].

DEFINITION 4.4.– A timed synchronized PN is a couple $\langle R_s, \tau \rangle$ where:

– R_s is a synchronized PN: $R_s = \langle R, E, \mu \rangle$, $R = \langle P, T, w^-, w^+, M_0 \rangle$;

– τ is a set of increasing mappings $\{\tau_{P_i}; P_i \in P\}$ from $\mathbb{R}^+$ to $\mathbb{R}^+$, associated with the places of the PN. τ_{P_i} is the unavailability duration of tokens added in place P_i. The set $\mathbb{R}^+$ of positive or null real numbers is used as time measure.

The draw of a timed and synchronized PN is the same of the underlying autonomous PN with the indication, next to the transitions and places, of the respective mappings $\langle \mu(T_j) \rangle$ and τ_{P_i}.

As for timed PNs, a token may be available or unavailable. A transition T_k is enabled by a marking M in the timed and synchronized PN if and only if it is enabled by M_a in the underlying synchronized PN. The firing of T_k at the time t extracts available tokens in $I(T_k)$ and adds in $P_j \in O(T_k)$ available tokens if $\tau_{P_i}(t) = t$ or unavailable tokens if $\tau_{P_i}(t) > t$. The application set τ then manages explicitly the time. Complete evolution rules may be found in the previously cited references and in the books from David and Alla [DAV 89, DAV 92].

4.5. Interpreted PN

An automated system is generally considered as a combination of two parts: the operative part and the control part. The operative part is modeled by a set of variables and operators. At a given time, the variables represent the state of the operative part; they may be modified by the activation of an operator. The control part is responsible for the scheduling of the operative part operators. It is an automaton whose outputs activate the operators and whose inputs receive information of the operative part. In this context, Moalla introduced the interpreted PNs [MOA 85].

DEFINITION 4.5.– An interpreted PN is defined by:

– an operative subsystem $\{V, OP, C\}$ with:

- V : $\{v_1 \ldots v_m\}$ a set of variables with values in the domains $D_1, D_2, \ldots D_m$;

- OP : $\{Op_1 \ldots Op_n\}$ a set of operators: internal mapping in $D_1 \times D_2 \times \ldots \times D_m$;

- C : $\{C1 \ldots Cr\}$ a set of conditions (or predicates) on the variables.

– a timed and synchronized PN $\langle R, E, \mu, \tau \rangle$ with $R = \langle P, T, w^-, w^+, M_0 \rangle$;

– an application $\varphi : P \rightarrow OP$;

– an application $\psi : T \to C$.

On the draw of the underlying timed and synchronized PN, the operator $\varphi(Pi)$ and the time function τ_{P_i} are indicated next to each place and the event $\mu(T_{ij})$ and predicate $\psi(T_{ij})$ are indicated next to each transition.

The marking evolution includes the rules of timed and synchronized PNs but requires in addition that a transition T_{ij} is receptive to a set of events E at time t only if:

– T_{ij} is enabled by the marking;

– $\mu(T_{ij}) \in E$;

– $\psi(T_{ij})$ is satisfied for the values at time t of the variables of the operative subsystem.

The main interest of this formalism is the ability to model hybrid phenomena where continuous and discrete variables interact together. It has been used as support model for the conception of systems including dependability requirements [AUB 91, JAM 01, MOA 80, AUB 87, JAM 00].

A special case of interpreted PN is called a control interpreted PN, including the following restrictions: it is safe (unicity of the token), not timed and deterministic. The variables of the operative part are modified as well by the process to be controlled (whose state is estimated through a set of sensors) as by the operators associated with the places of the PN. The transitions are synchronized by events of different origins (internal events, events coming from the process or from the human operator, sampling clock events, etc.).

4.6. Colored PN

4.6.1. *Introduction example*

Let us consider a machine park composed of two types denoted by a and b. Each machine type may have two states: in operation or under repair. The behavior of each machine type is represented by the PN of

Figures 4.3(a) and (b) with the two states (P_1 and P_2 for machines of type b, P'_1 and P'_2 for machines of type a) separated by the transitions (T_1 and T_2 for machines of type b, T'_1 and T'_2 for machines of type a) labeled by the failure (d_a, d_b) events and end of repair (r_a, r_b) events. The places P_1 and P'_1 receive the initial marking indicating the number of machine of each type (in the figure, only one token is present).

The PN of Figure 4.3(c) compresses these two PNs in one where a "color" is attached to the tokens to distinguish the type of machine a or b. In P_{11} a token b and a token a indicate that both machines are in operation. When a machine fails, a firing of transition T_{11} is operated extracting a token of the concerned color (a or b) from P_{11} and putting down a token of same type in P_{22}. It can be seen in the figure that the arcs are labeled by functions. In Figure 4.3(c), the function identity is associated with the arcs indicating that the token to be extracted is of the same type. This means that the enabling transition process and the concept of firing are completely independent, the firing of a transition is not a token flow. This will be illustrated in continuing the example.

Each PN of type a or b may also be folded on itself as shown by Figures 4.3(d) and 4.3(e). On the first one, a token of type O (operation) is in the unique place. If it fails, a token is extracted of this place (identity operation) but a token of type R is set down in the place by the function $f(O) = R$ indicating that the machine is now under repair.

The notion of weighting of the arcs is then replaced by functions associated with the arcs allowing to make a change (or combination) of colors and the numbers of the tokens. In addition, it is necessary to specify for each transition the type of color to which it is sensitive. In Figure 4.3(f), the transition is sensitive to all token types.

In our example, it is possible to merge the two PNs of Figures 4.3(d) and 4.3(e) to obtain the PN of Figure 4.3(f). The token colors are then combinations of $< machine\_state, machine\_type >$ ($< O, b >, < R, b >, < O, a >, < R, a >$). At each firing of the transition, it is necessary to define the color of the token to be set down. For example, $f(< O, b >) =< R, b >$ means that if the token defined by Id is of type $< O, b >$, then the color of the token to be set down is $< R, b >$. In this last PN, the transition is sensible to all of the token types.

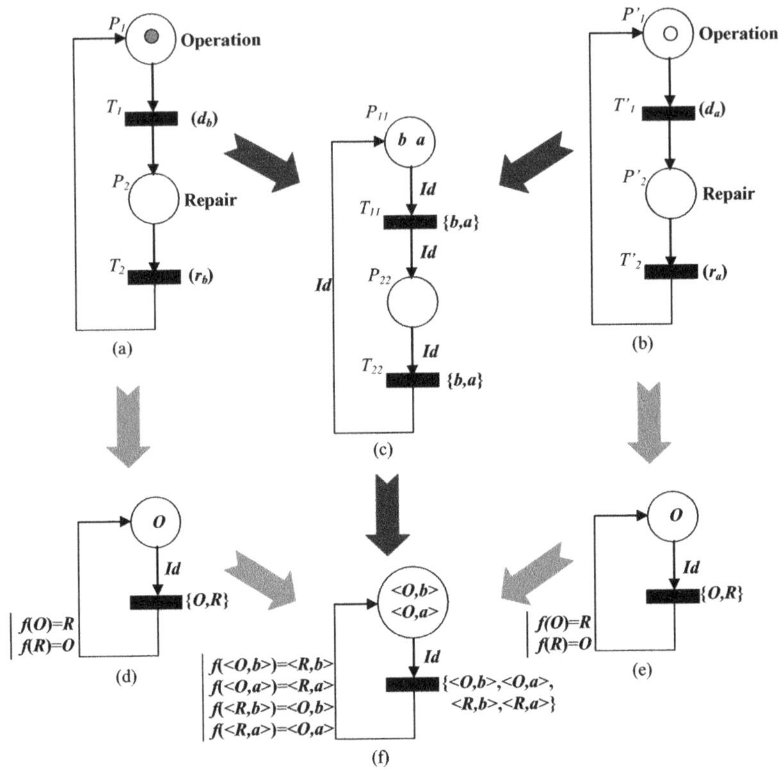

Figure 4.3. *Introductive example of colored PN*

This example gives an idea of the reduction power of a PN model due to the coloring concept. A single place and a single transition are sufficient to model the state of a park of n machines of different types. Practically, the replacement of weights associated with the arcs by functions operating on colors of different possible types gives a high expression power to colored Petri nets (CPNs). This has been formally formulated and completed by Jensen *et al.* [JEN 97a, JEN 97b, JEN 95, JEN 09].

4.6.2. *Formal definition*

DEFINITION 4.6.– A CPN is a nonuplet

$$CPN = \langle P, T, A, \Sigma, V, C, G, E, I \rangle,$$

where:

– P is a finite set of *places*.

– T is a finite set of *transitions*, $P \cap T = \emptyset$.

– $A \subseteq P \times T \cup T \times P$ is a set of directed *arcs*.

– Σ is a finite set of non-empty *color sets*.

– V is a finite set of *typed variables* such that $Type[\nu] \in \Sigma$ for all variables $\nu \in V$.

– $C : P \to \Sigma$ is a *color set function* that assigns a color set to each place.

– $G : T \to EXPR_V$ is a *guard function* that assigns a guard condition to each transition *t* such that $Type[G(t)] = Bool$, the notation $Bool$ standing for Boolean data type (where $EXPR_V$ is a mathematical expression defined on the typed variables of set V).

– $P_r : T \to \mathbb{N}^+$ is a priority function that assigns a priority to each transition.

– $E : A \to EXPR_V$ is an *arc expression function* that assigns an arc expression to each arc a such that $Type[E(a)] = Type[C(p)]$, where p is the place connected to the arc a.

– $I : P \to EXPR_\emptyset$ is an *initialization function* that assigns an initialization expression to each place p such that $Type[I(p)] = Type[C(p)]$.

Tokens can have different colors representing data types (colors), e.g. Boolean, integer, string or more complex data structure. The arc expression functions define the number and the values of tokens removed or added in the upstream and downstream places of a transition. An enabled (by marking) transition will be fired only if the associated guard function is evaluated to true. For a given marking, when several transitions are enabled, the transition with the highest priority is fired.

4.6.3. *A dedicated software CPN Tools*

Originally developed by the CPN Group at Aarhus University (Denmark), CPN Tools is now steadily improved by the AIS group, Eindhoven University of Technology (Netherlands). Its main architects are K. Jensen, S. Christensen, L.-M. Kristensen, and M. Westergaard. According to the CPN Tools Website [AIS 15], "CPN Tools is a tool for editing, simulating, and analyzing Coloured Petri nets. The tool features incremental syntax checking and code generation, which take place while a net is being constructed. A fast simulator efficiently handles non timed and timed nets. Full and partial state spaces can be generated and analyzed, and a standard state space report contains information, such as boundedness properties and liveness properties".

An example of CPN is shown in Figure 4.4. This CPN models a system with six components of two types (two components of type c_1 and four components of type c_2). These components are in "working" state until a failure occurs (firing of "failure" transition). Two types of repairers exist in this system: the r_1 repairer type can repair only c_1 component type and the r_2 repairer type can repair only c_2 component type. This association repairer-component is modeled by the color set (r_i, c_i) assigned to the "repairmen on standby" place. The initial marking of this place shows that one repairer r_1 and two repairers r_2 are available. When a component is in the *fail* state and the corresponding repairer is on standby (this association is guaranteed by the same value of variable x on the output arcs from these states), the transition "startrepair" occurs. At the end of repair operation, the component returns in the *working* state and the repairer returns in the standby state. The main interest to use CPN tools is the reduced size of the model. By comparison, a classical PN must have a number of places and transitions two times larger to model the same system.

Additionally, to the color concept, the probabilistic dependability assessment requires time management [JEN 97a]. The current time is given by a global clock. In addition to their color, the tokens contain a time value, also called time stamp. When a transition is enabled, it is fired and changes the time stamps of tokens which are deposited in its output places. In these places, the tokens remain frozen and cannot be used to enable other transitions until the current model time (given by the global clock) is smaller than their time stamps. As soon as the time stamp of the tokens is greater than or equal to the current time model, these tokens can enable other transitions which are instantly fired. In other words, the time stamp describes the "earliest" model time from which a token can be used. In the CPN of Figure 4.4, this is modeled by the expressions of two input arcs of "Working" and "Repair" places. The expression "x@+floor(exponential(1.0/(!lambda)+0.5))" changes the time stamp of the token deposited in the "Working" place by adding the value obtained by the exponential distribution law with rate lambda. This value represents the duration before the occurrence of the next failure, when token sojourns in the "Working" place for this duration. In the same way, the duration of reparation is obtained by the expression "(y,x)@+floor(exponential (1.0/(!mu)+0.5))", and the token sojourns in the "Repair" place for this duration.

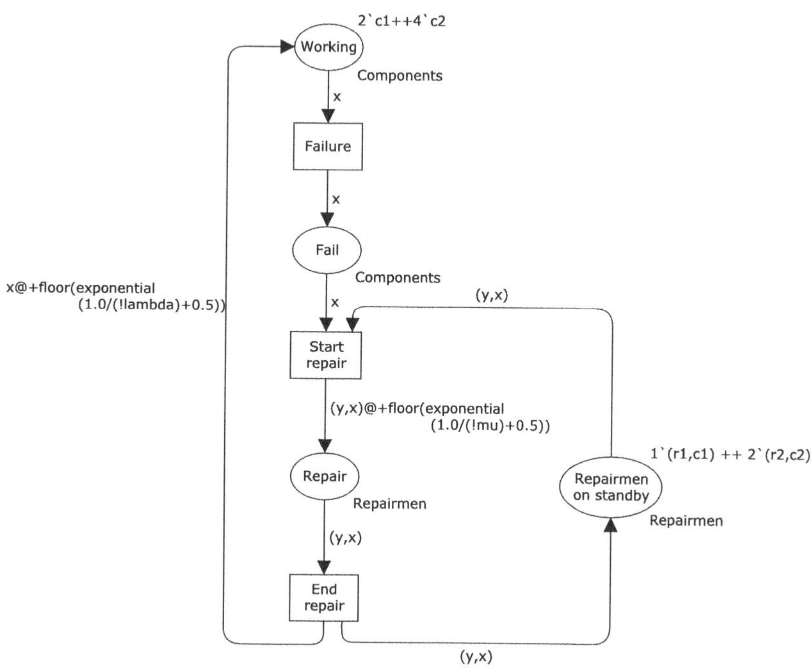

Figure 4.4. *An example of CPN Tools model*

For a better understanding, let us signal some specificities of CPN Tools:

– It gives the possibility of properties verification based on the analysis of the reachability graph which is modeled as a directed graph. It is then possible to answer a large set of questions concerning the behavior of the system, such as liveness, absence of blocking (deadlock or livelock), possibility to always be able to reach a given state (reachability), etc. This verification step is significant because these properties can represent, for example, specific safety properties of the modeled system.

– For a given marking, when several transitions are enabled, a priority mechanism is proposed. An integer is associated with each transition; more the integer is small, more the transition has priority. If all of the enabled transitions in a given marking have the same priority, then one of them is fired randomly.

– A transition can have an associated code region defining operations to be executed on its firing. It contains three fields: input, output and action defining the input and output variables and the algorithm to be executed.

Not illustrated in this example is the ability of CPN Tools to build a model with a hierarchical way as we will illustrate later (see Chapter 7). This concept allowing a modular modeling of complex system is implemented by the introduction of the "subpages" and "substitution transitions" concepts [JEN 97a JEN 09]. A subpage contains a PN describing a subfunction of the modeled system.

Individual CPN models (subpages) can be hierarchically related to each other in a formal way, i.e. with a well-defined semantics and formal analysis capabilities. CPN model hierarchy is realized through substitution transitions. The idea is to associate a transition (and its surrounding arcs) with a more complex CPN (a subpage), which gives a more precise and detailed description of the activity represented by the substitution transition. Figure 4.5 shows the CPN Tools hierarchical model.

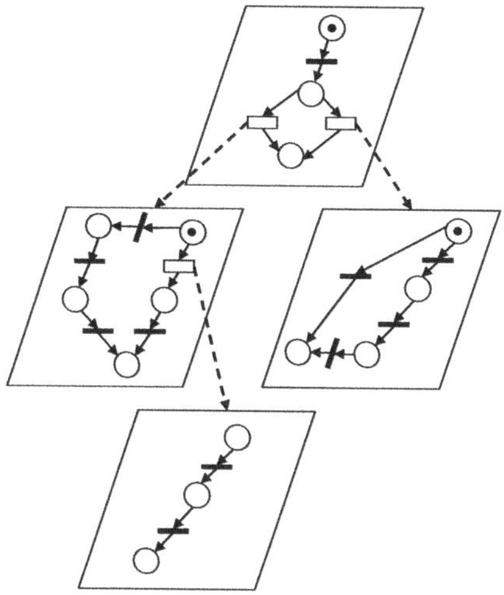

Figure 4.5. *Hierarchy in CPN Tools*

The places connected to a substitution transition, called "socket places", have clearly defined corresponding places, called "port places", in the related CPN module. They can transmit a given marking from a high level (level of substitution transition) to a low level (level of module) and vice versa. The number of levels in a hierarchical CPN is not limited, because a CPN module corresponding to a substitution transition can also contain other substitution transitions that are related to lower level CPN modules. A PN described in a module already used as a subpage may be assigned to another substitution transition. A lot of editing tools are included, for example the possibility to create new modules by the instantiation of an existing one. Any change in the generic model is immediately applied to all instances.

We will give examples of hierarchical models using CPN Tools in section 6.5.

Conclusion to Part 1

In this part, we made a short review of Petri nets (PNs) starting from autonomous PNs and introducing labeled ones to compare their expression power with regard to finite state automata presented in the first book of this series [AUB 15]. We saw that this power expression is greater due to the marking concept allowing, for example, to simply implement the need of memory. Although the PN models are simpler than finite state automata (FSA) models, the desire to find those which are even simpler resulted in the definition of colored PNs and other high-level PNs by the exploitation of symmetries or repeated parts in the graphs. However, we do not have to forget that the graph reduction is compensated by an increase in the associated code.

These basic models allow a representation of the scheduling of the various events in a given system without time considerations. For this, timed PNs were introduced. It is nevertheless important to remember that this introduction can induce some reductions of the initial properties of the underlying autonomous PN, particularly the state number of the reachability graph.

As many physical systems are dependent on an external environment, the latter must be taken it into account in the global model. Synchronized PNs were first introduced to consider the occurrence of events coming from this environment. Time was then considered in synchronized-timed PNs and finally the ability to handle variables representative of the environment evolution was given to interpreted PNs. This type of model has opened the door to hybrid modeling which will be used for dynamic reliability resolution problems as we will see later.

We did not make hypothesis on the nature of the events previously considered. For dependability studies, we have to consider failures and repairs of system components whose occurrences are, as everybody knows, characterized by probabilistic distributions. Their consideration is the subject of the Part 3.

We completed this first part with the introduction of an interesting piece of open software that will be used for the simulation of some systems presented later. Although it does not rely on stochastic PNs, CPN Tools allows the manipulation of random events and we will take advantage of this ability in the following part for the modeling and the evaluation by simulation of accidental processes in risk analysis.

PART 2

A Formal Approach to Risk Assessment

Introduction to Part 2

In the preface of the first book in this series [AUB 15], we defined the field of our studies. We said that although the dependability concept does not strictly include safety, there is no real border between them. For example, one of the objectives of the functional safety is the assessment of the availability of an electrical or electronic system dedicated to safety actions and the used methods and tools are the same. Recommendations of dedicated standards appeal for methods as formal as possible, even for the requested prior risk assessment. A risk assessment is the study of the occurrence of hazardous events and their consequences, more accurately their frequency and the severity of the caused damages.

Many publications, books and software tools are dedicated to risk analysis. Introducing an additional method is not our purpose, we will just propose to take the best of what exists to deduce a generic approach and introduce rigorous modeling particularly by the use of the Petri nets, formalism which has proven its efficiency in automatic control and dependability fields. This contribution does not lie at the preliminary analysis level, it supposes that risks have been beforehand identified. Thus, its objective is to move toward a generic approach of risk assessment that any industrial project actor could appropriate and to allow an easy nesting of elementary contributions to get the whole system risk assessment file. Any specialist should find in it a structure and some concepts compatible with its own practice.

To achieve this goal, we propose to move toward an ontology of the risk defined on the principle of a distinction between hazard supplier entity and vulnerable target entity, hazard and vulnerability spaces, entity states and events responsible for state changes. The specificity of this ontological approach relies on the fact that every concept is seen from two angles:

1) according to its semantic aspect;

2) regarding to its contribution to the accidental process.

This allows a systematic modeling of any process of hazardous evolution in a system containing one or more risks.

The definition of an accidental process as a state sequence and the proposal of a typology of events causing state changes of this process gives an efficient help to the conception of primary (protection) and secondary (prevention) safety barriers in charge of hindering the effects of these events. The state/transition modeling and especially the Petri net formalism allows us in addition to encounter the kinetics of accidental scenarios.

Ontology concept

The ontologies today occupy a pivotal position in several engineering domains. Indeed, in artificial intelligence, the semantic Web or software engineering, the need for information modeling through ontologies has increased so as to become a strong key-issue for a successful project.

Considering the interest it represents, the ontology has been tackled in some standards, for example in the ISO 21127 [ISO 14]. This standard describes in particular the metadata necessary to the ontology modeling.

The ontology is generally used for knowledge-based reasoning in a specific domain. In fact, the ontology itself constitutes a data-based model representing a set of concepts and their relationship.

However, before addressing the ontology in detail, it is necessary to dispel nuance that can exist with "terminology". In a terminology, we deal with words and meaning, i.e. the relationship between these words; while in an ontology, we focused on the notion of concept and relationships between concepts.

Ontology is an ancient Greek philosophical concept (Aristotle) described by Descartes and Kant as the "science foundation of knowledge". It has also been considered (17th Century) as "the part of metaphysics that applies to the being as being, regardless of its particular determinations". More recently, it has appeared as the "study of being as being, of being in itself". It is also the study of life in general, in existentialism.

In the domain of knowledge engineering, the concept of ontology was introduced and readjusted in the 1990s by Gruber *et al.* [GRU 93] "An ontology is an explicit formal specification (understandable by machines as well as humans) of a shared conceptualization (described consensual knowledge)". Generally, ontology describes individuals, classes (sets, collections and objects), attributes (properties, functions, characteristics and parameters), relationships between objects and events triggering changes on attributes or relationships.

Computer ontologies are tools that allow representing a specific knowledge corpus in a form usable by a machine. An ontology contains a hierarchical description of given domain concepts and a description of the properties of each concept and its relations with the others. The concepts are organized in a graph whose relationships can be semantic or subsumption. The "subsumption" organizes the concepts by abstraction of common characteristics to achieve a hierarchy corresponding to a taxonomic organization of objects ROC 05.

Typology and design criteria

There are many types of ontologies and they were the subject of classification. It is not our purpose here to make a review but simply to show that we place ourselves in the context of domain ontologies that can be seen as a specialization of generic ontologies.

Building an ontology is a process requiring method and compliance with some design criteria. Grüber [GRU 93] proposes five design criteria for ontologies:

1) Clarity. Definitions should be objective and complete. Formalism is a means to this end. When a definition can be stated in logical axioms, it should be. All definitions should be documented with natural language.

2) Coherence. An ontology should exclude inferences that are inconsistent with the definitions. At the very least, the defining axioms should be logically consistent.

3) Extendibility. It must make the definition of new terms for special uses possible, based on the existing concepts, in a way that does not require the revision of the existing definitions.

4) Minimal encoding bias. The conceptualization should be specified at the knowledge level without depending on a particular symbol-level encoding. Knowledge-sharing agents may be implemented in different representation systems and styles of representation.

5) Minimal ontological commitment. An ontology should require the minimal ontological commitment sufficient to support the intended knowledge sharing activities. An ontology should make as few claims as possible about the world being modeled, allowing the parties committed to the ontology free to specialize and instantiate the ontology as needed. Since ontological commitment is based on consistent use of vocabulary, ontological commitment can be minimized by specifying the weakest theory (allowing the most models) and defining only the terms essential to the communication of knowledge consistent with that theory.

Finally, to meet the five criteria of Gruber, it is required to determine precisely: the ontological commitments, the definitions of high-level categories, the refinement process of these commitments and definitions, and then the specialization of these categories [GUA 98].

Conclusion

Ontology is not a goal in itself; it provides means to build other models and systems. The degree of formalization (implantation from conceptualization structures in a formal language) varies from the natural language (without primitive) characterized by circular definitions, unclear statements and unstable objects, to the formal language characterized by a semantic, constructed expressions with primitives, connectors, etc. [CHA 00].

In the light of the added-value of developing risk analysis ontology, we will proceed to migrate from a natural language into a formal language by introducing the concepts of entity, situation and event.

5

Ontology-based Accidental Process

5.1. Preliminary definitions

Since the emergence of the societal demand in terms of dependability and safety, specialists have struggled to reach a clear consensus over the questions of terminology and over all the engineering methodology.

Unfortunately, if we draw up a balance sheet of the continuous efforts given during decades, a rather complex picture results.

We can summarize the terminological divergence either in syntax or semantics, in the following points:

– association of the same term to different concepts;

– association of the same concept to different terms;

– nuance between different concepts;

– perfunctory definition of terms often individually defined and presented in alphabetical order as glossaries;

– confusion between the notions of state and transition (event);

– lack of typical accidental process modeling;

– excessive use of textual descriptions.

In order to make a rational presentation of the essential terms and concepts of the risk analysis, let us follow a deductive reasoning starting from a proposition of the risk concept definition. Such a reasoning could lead to an easy establishment of a semantic diagram of the risk vocabulary:

1) *Risk:* set of the potential consequences for an *entity (2)* of being exposed to a *hazard (3).*

This definition introduces the two concepts of *entity* and *hazard* that must be defined in turn and so on:

2) *Entity:* the standard IEC 50 (191) defines the concept of entity (or item) as follows [IEC 90]: "Any part, component, device, subsystem, functional unit, equipment or system that can be individually considered". It may be natural or created by man including hardware, software or both, and may also in particular cases, include people. This definition clearly dependability oriented may be extended in the field of risk analysis to consider people, set of people, human organizations and even natural environment, as whole entities. The concept of entity seems to be more suitable as this of system used in the risk analysis Mads-Mosar method [PER 07] because it is more encompassing and so able to support a hierarchical description.

3) *Hazard:* any situation generated by an entity able to cause a potential *damage (4)* to another entity if a *hazardous phenomenon (5)* occurs. This definition although currently accepted is not always so clearly defined.

From this definition appears the need to distinguish two types of entities depending on whether they are hazard source or target. These two types of entities will be defined in the next section.

4) *Damage:* this generic term may take different forms depending on the nature of the entity that undergoes:

- human: physical or moral injury, harm, disability, death, etc.;
- equipment and goods: destruction, wear, loss of performance, etc.;
- organization (company, society, etc.): disturbance, threats, strike, bankruptcy, etc.;
- environment: air, water and soil pollution, radiation, etc.

A damage is generally expressed in terms of gravity (or severity). The gravity can be evaluated with numeral indicators (numbers of injured or killed people, financial cost, etc.) or estimated according to a predefined scale of gravity levels.

5) *Hazardous phenomenon:* event or sequence of events whose occurrence has an *accident (6)* as a consequence.

This definition leads to further clarify the nature of these events as they relate to source and target entities. These different types of events and their sequencing will be defined in section 5.3.

6) *Accident:* the production of an effective *damage (4)*.

These different definitions show the risk analysis as the research of the involved entities, their successive states and the events associated with these state changes. A state/transition model appears then suitable for the representation of an accidental process. With this in mind, let us specify the entities types, their successive states and the events associated with these state changes.

5.2. Elementary entities: HSE and VTE

5.2.1. *Hazard supplier entity (HSE)*

An HSE is an entity likely able to carry or generate one or several hazards under given circumstances. It may be, for example, a natural system as well as a system or a provision created by man.

In the field of technological risks, an HSE corresponds to a technical unit necessary for an expected operation in a given process.

The *hazard space* is the space in which an HSE induces one or many hazards. This space may be multi-dimensional, inside and/or around the HSE, temporarily or permanent.

5.2.2. *Vulnerable target entity (VTE)*

A VTE is any entity likely to be a victim of damage when exposed in a hazard space under given circumstances. It may be, for example, a person, a good or an environmental component, etc.

The *vulnerability* characterizes the potentiality of a VTE, when exposed in a hazard space, of being able to undergo given damage.

The *vulnerability space* of a VTE is characterized by the weakness or the absence of devices dedicated to the damage avoiding or to its gravity reduction. It may also be multi-dimensional, temporarily or permanent.

5.3. Elementary situations and elementary events

5.3.1. *State versus situation*

We choose Petri nets (PNs) as a model for the representation of the accidental process in a system where entities are interacting. As a result, we must reserve the term "state" to qualify the marking vector of the PN modeling this process. To characterize the state of an elementary entity, we will reserve the term "situation".

An event is the dual concept of state of an entity; in other words, it can produce or be produced by a change of this state. An event is characterized by its occurrence (date, frequency, number, etc.) and can be deterministic or random (stochastic). In our analysis, events are responsible for situation changes so we will describe them alternatively.

5.3.2. *Initial situation (IS)*

This is the situation considered as normal where any entity belongs to its functional specification before the development of any accidental process. In this situation, no VTEs are exposed to a hazard and any potential HSE does not radiate a hazard.

5.3.3. *Initiating event (IEv)*

This is an event, internal or external to the system, having the capacity to induce a situation change of an HSE from the initial situation. It initiates a hazard process.

5.3.4. *Hazard situation (HS)*

This is a situation of an HSE after the occurrence of an initiating event. The hazard space is then materialized (dispersion of a pollutant gas cloud, skid of a car, etc.).

5.3.5. *Exposure event (EEv)*

This is an event occurring in the initiating situation of a VTE, internal or external to the system, which induces the exposition of a VTE to a potential hazard space.

5.3.6. *Exposure situation (ES)*

This is a situation of a VTE after the occurrence of an exposure event.

A passenger having penetrated (exposure event) on the rail track is regarded as a VTE in potential exposure to several hazards (electrocution if the track is electrically powered, crushing if a train enters the station, legal proceeding if a control agent is present, etc.). A scientist entering a radiation effect zone in a laboratory becomes an exposed VTE.

5.3.7. *Accident situation*

This situation is the logical consequence of the intersection of the vulnerability space of one (or more) VTE with the hazard space of one (or more) HSE.

5.3.8. *Hazardous (feared) event (HEv)*

The accident situation can be reached as soon as a both vulnerability and hazard spaces interfere. It is the case when people without respiratory mask are present in a closed space where a leakage of lethal gas occurs in a pipe. However, in other cases, the occurrence of an event may be necessary to cause an accident situation. In the previous example, if we are dealing with an explosive gas, the leakage must be followed by an ignition phenomenon to produce the accident situation. This is the type of event that we will designate as a hazardous or feared event.

5.4. Conclusion

Although it may appear unsophisticated, this enumeration is still effective in the purpose of risk assessment as we will see. Indeed, it will allow the description of elementary accidental processes as well as feared sequences of accidental processes considering that the feared event of a process may become initiating or exposition event of another process. To demonstrate this, a formalism must be used and it must be able to support assessment tools as quantitatively as possible. This is the purpose of Chapter 6.

6

Petri Net Modeling of the Accidental Process

As we have just seen, entities can be in different situations all along the evolution of an accidental process evolving to the will of events. Resorting to Petri nets (PNs) as a modeling support of this state transition alternation then comes naturally. Recently, a PN-based model of risk assessment has been proposed by Grunt and Bris [GRU 15] for modeling of the risk to safety of personnel on an offshore hydrocarbon production facility following hydrocarbon. The advantages of such a dynamic model are emphasized compared to the steady state approach of event tree models. However, the authors do not propose a generic modeling approach as we will develop in the following sections.

We will then formalize under the form of a PN a generic accidental process summarizing the chaining of the state-transitions alternations previously described.

The initial situation of elementary entities (Hazard Source Entity (HSE) and Vulnerable Target Entity (VTE)) will be represented by initial places in the PN model. They will then be marked at the initial time, that is to say the beginning of the system lifetime, for example. The state of this entity can only evolve on the occurrence of an exposure event for a VTE or an initiating event for an HSE. This is represented in Figure 6.1 by the transitions to which these events are associated.

REMARK 8.1.– Formally, it is a matter of a particular PN class. In case the research of possible event sequences is the main interest, the labeled PN class may be selected. If time is the main interest, the timed and/or synchronized PN classes may be suitable. If all events may be characterized by a constant occurrence rate (exponential distribution of the occurrence times), the class of stochastic PN (defined in Part III) is a must.

As of course an HSE or a VTE may be associated with several hazards, it is possible to define as many output transitions to the initial places of the model, each of them being associated with a specific event. A given HSE can then be associated with many VTEs. In this case, requirement of colored PN model may be useful. It becomes necessary for the initial place associated with an HSE to contain one token for each of the possible output transition types. Indeed, the firing of the transition relative to the initiating event associated with the first hazard does not necessarily induce the potentiality loss of the other hazards.

6.1. Elementary process

An elementary accidental process is presented in Figure 6.1 where the state of the system at the initial time instant is represented by the initial marking. The situation evolution of the HSE and VTE from their initial situation (places P_1 and P_2) to the exposition situation (place P_3) or the dangerous situation (place P_4) is done by firing of transitions T_1 or T_2 on the occurrence of exposition or initiating events. When both are fired, a new system state is reached where P_3 and P_4 are marked. However, in this state, the overlapping of the hazard and exposition spaces of the two concerned entities can be spontaneous or evolve until the occurrence in this common space of an event: the feared event. In both cases, the transition firing of T_3 is associated with this event. In the first case, it is inherent to the source/target couple and may be considered as a produced event by the immediate firing of T_3, in the second case, it is a received event, acting as a supplementary condition for the firing of T_3. When T_3 is fired, P_5 becomes marked indicating that the HSE/VTE couple is in accident situation.

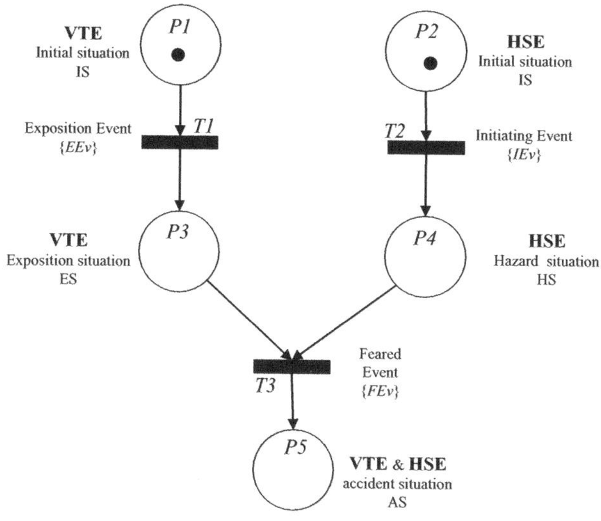

Figure 6.1. *Synchronized PN of an elementary accidental process*

Let us give the example of the risk for a handling worker being injured by the traveling crane of a warehouse. The traveling crane is the hazard source. The initiating event is the hooking of a load by the winch and the hazard space is the part of the warehouse under the winch. The exposition space of the workman, which is the target, is its own physical space and the exposition event is the penetration under the load of the traveling crane. A feared event may be the break of the carrier cable of the load.

Another example is the risk for a worker of being asphyxiated in an industrial premise in which a chemical reactor can release a toxic gas. The source is the pressure relief valve of the reactor, the initiating event is the opening of the valve and the hazard space is the whole premise. The target is the workman in the premise who has the obligation to wear a breathing mask. The exposition event is the removing of the mask and its exposition space is his/her respiratory system. The accident situation is here obtained as soon as both hazard and exposition situations are present. The feared event is precisely the firing time instant of transition T_3.

A finite state automaton-based model could have been suitable to represent an accidental process. However, it seems to be more interesting to resort to PNs because the model may be progressively built without an *a priori* knowledge of all the set of the potentially accessible states in a complex process. So, the necessary conditions to the initiation of a accidental process are materialized by the marking on the PN model. At least one HSE and one VTE are requested.

The model of Figure 6.1 may of course be completed to consider some particular conditions. First, let us consider the case where the exposition situation (place P_3) and the dangerous situation (place P_4) have a finite duration in time. It may be possible that the feared event does not occur before the events created by these situations. It is then necessary to remove the token from places P_3 or P_4. This may be achieved by introducing the transitions T_1' or T_2' associated with the corresponding events "end of exposition" $(/EEv)$ and "end of initiation" $(/IEv)$. The corresponding PN is shown in Figure 6.2.

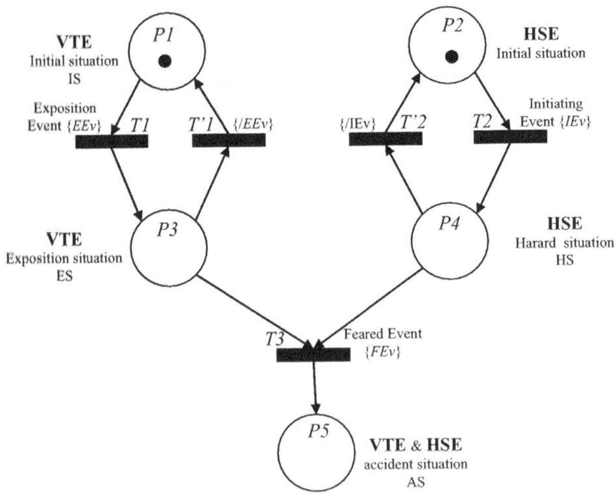

Figure 6.2. *Completed elementary accidental process*

Rarer is the case where an entity is a hazard supplier for itself. The same entity is both an HSE and a VTE and a single event initiates the firing of T_1 and T_2. This may be represented, for example, by merging T_1 and T_2 as well as P_3 and P_4.

Enriching the model to take account, for example, of the insertion of safety barriers is possible by the introduction of other places in the PN model which will impose, through their marking, new conditions on the transitions firing besides the associating events.

Due to the concepts of marking and arc weight, it is also possible to easily model some more complex phenomena such as a situation evolution related to the accumulation of a parameter.

Another advantage of PNs is the ability to build from separated elementary accidental processes the scenarios leading to a domino effect where several HSEs or VTEs interact sequentially together.

Such mechanisms will be detailed later.

6.2. Sequence of elementary processes

Several elementary processes can be chained in a scenario, sometimes called a domino effect. This is the case where the feared event of an elementary process acts as the initiating or the exposition event of another elementary process. It may be considered that the feared event is emitted by the first process and received for synchronization by the second. Figure 6.3 shows such a chain of four elementary events. It is then possible to make the correspondence between the different events: $FEv1 = EEv3$, $FEv2 = IEv3$ and $FEv3 = IEv4$.

6.3. Modeling the action of a safety barrier

A safety barrier is not a specific entity; it is dedicated to the reduction of the occurrence probability of an event or the severity of a damage. It is an entity like any other in the system which can itself be a vulnerable target (failure, destruction, malevolence, etc.), or even a hazard supplier if the impact of its introduction in the system has not been seriously studied.

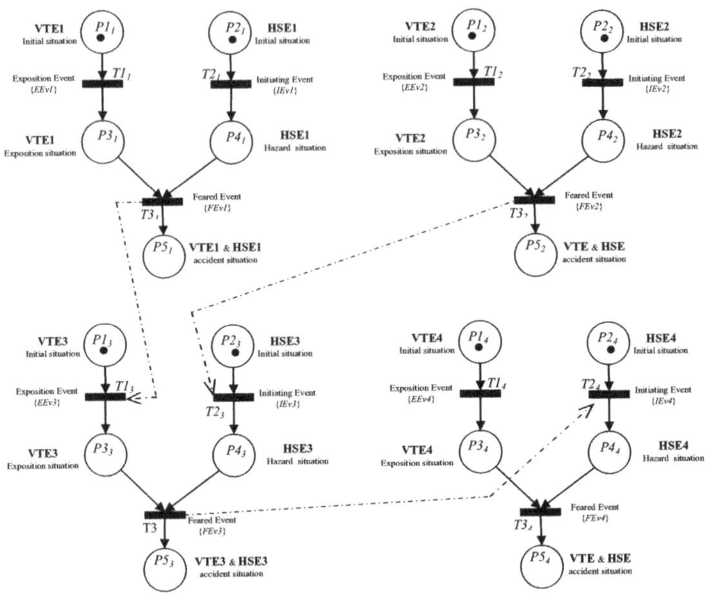

Figure 6.3. *Chain of elementary processes*

In the PN model of an elementary process, we will then introduce the part of a safety barrier as a supplementary condition to the firing of a given transition. So, the presence of the event labeling the transition will not be sufficient for it to be fired, it will be necessary that an additional condition be satisfied under the mean of the marking. This is shown in Figure 6.4.

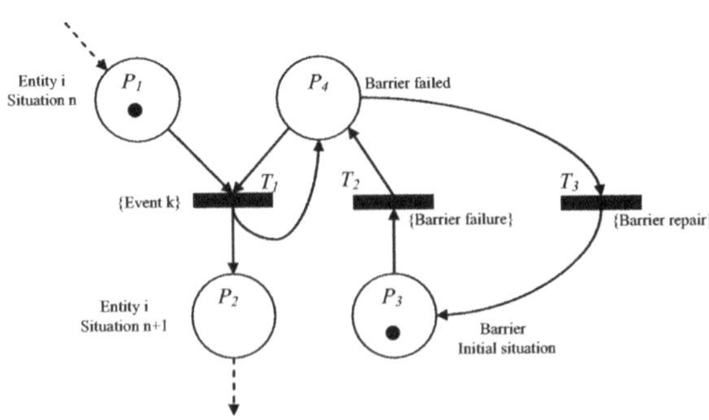

Figure 6.4. *Action model of a protection barrier*

The initial marking of the PN concerns places P_1 and P_3, meaning that the entity i is in the situation n and that the safety barrier is active (initial situation). It appears that transition T_1 cannot be fired because place P_4 is not marked meaning that the safety barrier is not failing. If unfortunately the safety barrier failure occurs (T_2 fired), then the token of P_3 moves to P_4 and the transition T_1 becomes fireable. It will be fired as soon as the event k occurs, meaning that the entity moves from situation n to situation $n + 1$. However, the firing of T_1 consumes the tokens of places and P_4. This is satisfying for P_1 but not for P_4. The barrier remaining in fail state P_4 must stay marked and it is the role of the arc $\{T_1, P_4\}$ which redeposits a token in P_4 after the firing of T_1.

The repair action of the barrier is modeled by the transition T_3 allowing its locking action for a next eventual solicitation of T_1.

6.4. Modeling of a cumulative process

This type of process is not uncommon in a risk analysis context. It is, for example, the case of the monitoring of radiation doses received by operators in nuclear plants. At each penetration in the hazardous area, the operator receives an elementary radiation dose represented in the model by a token. The dose accumulation is represented by an accumulation of tokens in a place of the model. The weight of the arc leaving this place corresponds to the maximal acceptable number of doses. As soon as the number of token in the place becomes equal to the weight of the arc, the downstream transition becomes enabled and may be fired to make the operator in a new situation where he/she is no longer authorized to access the hazardous area.

In Figure 6.5, place P_1 represents the operator (VTE) in an initial situation. The exposure event (for example, the not wearing of protection against radiation) moves the operator into an exposition situation. When he/she enters the hazardous area (VTE in a dangerous situation), the transition T_2 is fired; he/she is then exposed to radiation and place P_4 becomes marked. Note that this new situation (new place in the PN) does not suppress the dangerous character of the HSE, this is why a token must be eventually redeposed in P_3 (dashed arc $\{T_2, P_3\}$).

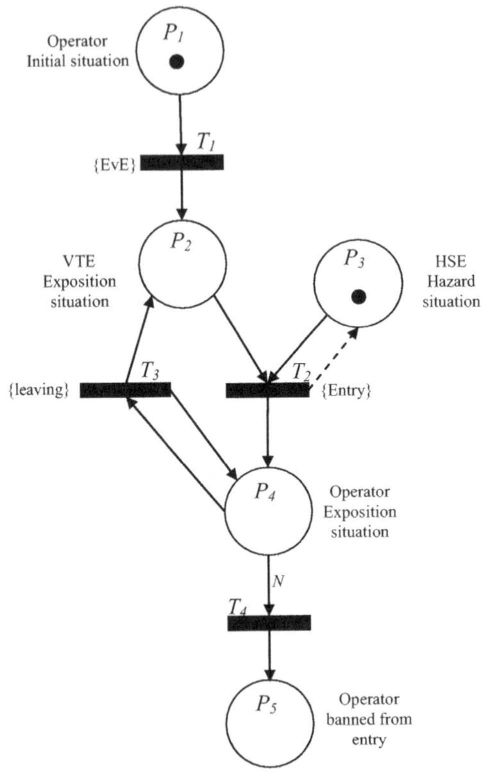

Figure 6.5. *Modeling of the cumulative process*

The transition T_3 is fired by the occurrence of the event "leaving of the hazardous area" and the operator goes back to the exposition situation. To materialize the fact that the operator retains his/her radiation doses, an arc $\{T_3, P_4\}$ is added to redepose the token consumed by the firing of T_3. So, the place P_4 accumulates the token until their number becomes equal to the admissible maximum N. The transition T_4 is then fired consuming the whole set of N tokens. Putting us a barrier against the return of the operator is representable by making the firing of T_2 conditional on the non-marking of P_5 (this may be made without difficulty).

6.5. PN as a support for risk assessment

Usually, the risk assessment of a system consists of the accumulation, with an eventual expansion coefficient, of the different risks associated with the accidents potentially occurring in this system. As showed in the previous section, the causes of these potential accidents may be correlated, for example, in the case of domino effect. Estimating these correlations by an *a priori* expansion coefficient seems to be a dangerous approximation. The PN modeling of the accidental process may be an efficient way to avoid such approximation types provided that some additional attributes are introduced to model the damage level of the potential accidents and the frequency of the various events responsible for the evolution.

6.5.1. *Modeling of the damage*

Each place P_n corresponding to a potential accident is associated with an attribute $D(P_n)$ which describes the damage. This attribute may be simply an integer or a weight according to the adopted codification. In the case of multiple damage types (technological, financial, social, human, etc.), each of them will be codified and the attribute becomes a vector.

6.5.2. *Modeling of the event frequencies*

The various events responsible for the accidental process are associated with transitions of the model. Although these events are generally of random type, they are rarely modeled by stochastic variables in risk analysis as it is the case in reliability assessment. Common practices in risk analysis associate with events a value of their mean occurrence frequency such as once a month, once a year, once a decade and so on. The introduction of the mean occurrence frequency concept in a PN model is not a trivial work. We can, for example, imagine it as a token color to which a special algebra is associated. Instead of a substantial development of such an algebra, we thought that it is much simpler to resort to simulation with a periodical generation of the various “primary” events, that is to say events not coming from another PN (see sequence of elementary processes).

We will then build event generators (see Figure 6.6) to periodically produce the necessary events. On the simulation duration, tokens will accumulate in the final places of the PN describing the hazardous process. The obtained number divided by the simulation duration will give a mean value of the occurrence frequency of the accident. To define the simulation duration, it will be necessary to deposit as many tokens as needed, in the places corresponding to initial situations of the entities considering the respective frequencies of the various initiating and exposition events. Another possibility is to consider the initial situations as infinite token sources and stop the simulation after a predefined time.

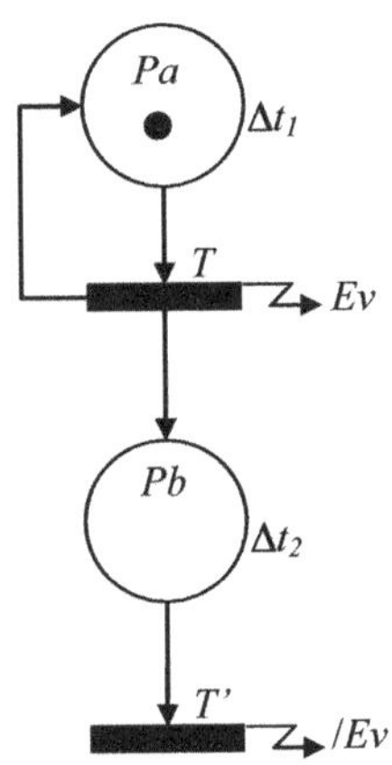

Figure 6.6. *PN model of the event generator (mean values)*

Figure 6.6 shows a simple generator of two generic complementary events Ev and $/Ev$ (for example, exposition event and end of exposition event). It is a P-timed PN, with outputs to produce events and whose the token in the initial place Pa is reserved for the time delay Δt_1 , mean period between two successive occurrences of event Ev. A token on the place Pa is used to determine when the new event arrives. After Δt_1, the firing of T occurs and the second place Pb becomes marked but its token remains reserved for the time $\Delta t_2 < \Delta t_1$ (to ensure the alternation of Ev and $/Ev$). After Δt_2, the firing of $T\prime$ generates the event $/Ev$. Let us remark that the firing of T redeposits a token in Pa in order to produce the next occurrence of the event Ev and so on. These two events are part of the environment of the synchronized PN describing the accidental process of the studied system.

The PN model of the accidental process may then be used for the risk assessment by simulation in addition to the ability of an exhaustive event sequence generation. It is of course important to precise that a strong hypothesis was implicitly admitted supposing that the real behavior of the process converges toward the "mean" behavior modeled by the PN simulation model. Nevertheless, it is possible to release a little this hypothesis by the introduction of a random character in the simulation.

6.5.3. *CPN Tools implementation*

The event generator shown in Figure 6.6 may be easily modified to introduce a random variation of the delays around their mean value.

The time delays are considered as random variables with normal distribution around a central value with a given standard deviation. Nevertheless, it is imperative to complain with the condition $\Delta t_2 < \Delta t_1$ on the currently drawn values and not only on the mean value. During the event generation, if the random draw does not produce an event $/Ev$ before the next event Ev , the value of Δt_2 must be limited to satisfy the above condition. In addition, as CPN Tools handles P-timed PNs, reserved tokens cannot be used for any transition validation and that is why places are inserted to indicate, when they are marked, that the time delays are in process.

Figure 6.7 is the implementation of such a generator with CPN Tools.

The places "next_generation_occurrence" and "next_disable_occurrence" implement the places Pa and, respectively, Pb in Figure 6.6. The transitions "event_generation" and "event_disabled" implement the transitions T and, respectively, $T^{'}$.

The main advantage of the model in Figure 6.7 consists of its generic character. It models the generation of all events types that must be considered in the whole system model. It is dynamically instantiated for each event. The event associated with an instance is given by the token color in the place “event_id”.

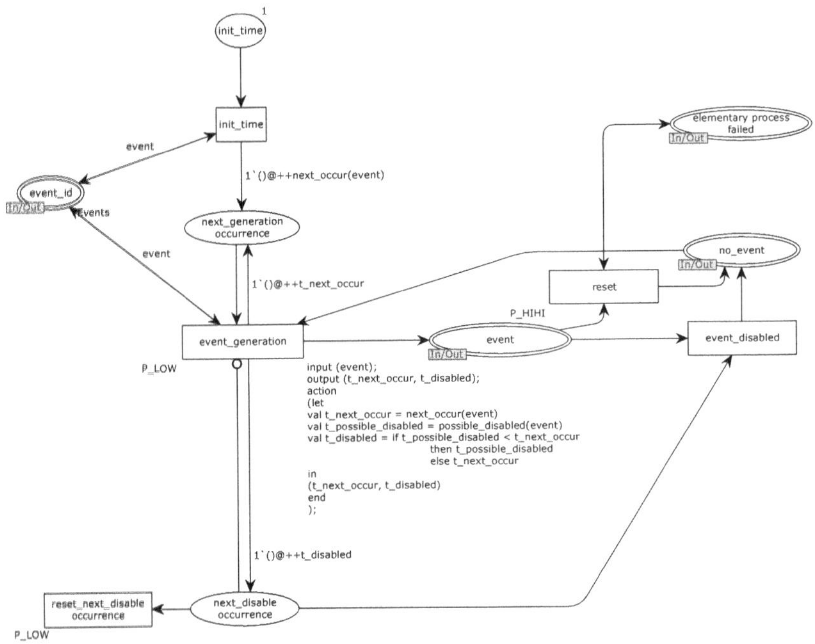

Figure 6.7. *Event generator CPN Tools model (stochastic values)*

Let us give some details on how this generator runs:

When the transition “event_generation” fires, a random value is drawn for the variable “t_next_occur”. This value represents delay to the next occurrence of the event Ev and it is used to determine the time (Δt_1) at which the next occurrence of this event will occur. The times (delays) between two occurrences are randomly distributed around a mean value. This delay is computed by the function “next_occur” according to the event type. It represents a first output of the code region of the transition “event_generation” and it is used to modify the time stamp of the new token that will be added in the place "next_generation_occurrence". The time stamp of this token determines when the next occurrence of the same event will appear (next firing of transition "event_generation)". There is no token on the place "next_generation_occurrence" in the initial marking for enabling transition "event_generation". The "init_time" transition is the only transition that is enabled in the initial marking (the place "init_time" contains one token given by inscription "1" in its top right side). This transition is used to put a token on the place "next_generation_occurrence". The delay computed by the function "next_occur(event)", in inscription on the arc from the transition "init_time" to the place "next_generation_occurrence", is used to ensure that the first occurrence of event Ev will not always occur at time 0 for different simulations.

Always when the transition "event_generation" fires, a random value is drawn for the variable "t_possible_disabled" that represents possible delay to the next occurrence of event $/Ev$. This random value is computed by the function "possible_disabled" according to type of event. The effective delay ("t_disabled") will be the minimum between the values of "t_next_occur" and "t_possible_disabled" in order to ensure compliance with the constraint $\Delta t_2 < \Delta t_1$. This delay is the second output of the code region of the transition "event_generation" and is used to determine for how long time the token added in the place "next_disable_occurrence" will be reserved (Δt_2).

The added tokens in places "next_generation_occurrence" and "next_disable_ occurrence" being timed are consequently reserved and cannot be used to enable any transition. To memorize this state, the place "event" is added to the PN indicating that the event has occurred until the transition "event_disabled" will be fired. In the same way, the place "no_event" memorizes the fact that the transition "event_disabled" has been fired until the next firing of transition "event_generation".

Moreover, to ensure that new firing of "event_generation" transition representing new event occurrence will be held only after the firing of transition "event_disabled", an inhibitor arc is used from the place "next_disable_occurrence" to transition "event_generation".

The double line bordered places are input/output places of the PN in Figure 6.7 representing the generator model of an event. This PN is a subnet of the global model for risk assessment that is hierarchical model according to Figure 4.5. The type of these places (in, out or in/out) is represented by rectangular port tags positioned next to them and specifying whether the port place is an input, output or input/output port. This supposes that a PN model at the upper level of hierarchy describes the relationships between the different parts of the whole model. We will choose as an example of an accidental process that links two elementary processes in which the feared event of the first one is the initiating event of the second. This is described by the high-level PN in Figure 6.5.4, where the substitution transitions (see section 4.6.3, double bordered line) represent sub-PNs describing a more detailed view of the behavior. In our modeling, these substitution transitions represent either an elementary accidental process or an elementary event generator. Moreover, the first elementary process has two input events: one of them representing its exposition event and the other representing its initiating event. Additionally, the second elementary process also requires a third event accounting for its exposition event. Each of the substitution transitions representing the three events is associated with one instance of the event generator PN (Figure 6.7) at the lower level of hierarchy. The substitution transitions representing the two elementary processes are associated with one instance of PN model of an elementary process. The implementation of elementary accidental process (Figure 6.1) with CPN Tools is given in Figure 6.9. The places connected to these substitution transitions (in this upper level) represent port input/output places of the generator and, respectively, elementary process PN models (Figures 6.7 and 6.5.4). These places are connected to the port places (in the lower level of hierarchy) by a socket-port mechanism (each socket in the upper level has its corresponding lower level port and they are identically marked anytime). The places "Exp_event_1", "Exp_event_2" and "Init_event" are colored places. Their initial marking gives, respectively, the identification of event that is modeled by the connected substitution transition. Each of these places is connected to the colored port place "event_id" in the corresponding PN instance of generator event model (Figure 6.7).

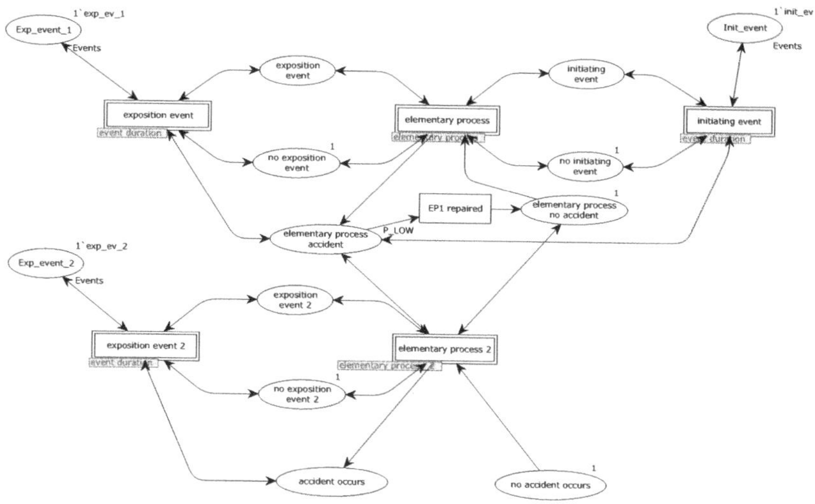

Figure 6.8. *Simulation model of a sequence of two elementary processes*

In addition to the model of elementary accidental process previously developed (Figure 6.1), the implementation model shown in Figure 6.9 contains four port places (double bordered line) that are input/output places. These places are connected to the places "event", respectively, "no_event" of corresponding PN of event generator model (Figure 6.7) by means of the socket places situated in the upper level (Figure 6.5.4) of the hierarchical model. Thus, these places and their connection with corresponding places (in other PNs) model the sending/receiving mechanism of events Ev and $/Ev$ associated with transitions in Figures 6.6 and 6.2 that represent fundamental models for the events' generating and elementary processes evolution. In order to assess the accident frequency, the following scenario is considered: the domino effect from the first to the second elementary process takes place only if the second process is already in the exposition situation when the feared event of the first accidental process occurred. Otherwise, the system returns in its initial situation (repairs and corrective measures to remove the feared event of the first elementary process are neglected; it goes here in a safety context when after the first accident the process is stopped). To model this scenario, following elements are added:

– output arcs from transition "Feared-event" to places "VTE_Initial_situati-on" and "HSE_Initial_situation" to reset the PN model of elementary process (Figure 6.9);

– transition "reset" that removes token from the place "event" and adds token in place "no_event" of the PN model of corresponding event generators to the elementary process that is in accident situation (Figure 6.7); this transition has the highest level of priority (P_HIHI) in order to remove the token of place "event" and to prevent that it can be used somewhere else in the PN because it is expired after the feared event and the reset procedure;

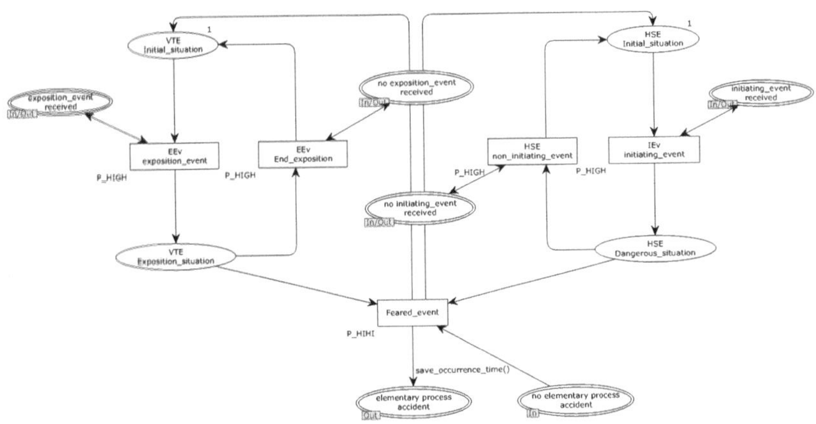

Figure 6.9. *Simulation model of the elementary process*

– to complete the reset procedure, the transition "reset_next_disable_occurrence" is necessary to remove the token from the place "next_disable_occurrence"; this transition has the lowest possible priority (P_LOW), in order to give priority to transition "event_disabled" (with medium priority that is the default priority of transitions, not indicating in the PN model) when the place "next_disable_occurrence" is marked; thus, the transition "reset_next_disable_occurrence" is enabled only after the previous firing of the transition "reset" that removes token from the place "no_event".

Thus, in Monte-Carlo simulation of the entire PN, the feared event of the first elementary process is reset inside a simulation story, if the conditions for the domino effect are not given. A story is stopped only when the feared event of the second elementary process (representing the accident occurrence of the whole process) has occurred.

6.5.4. *Evaluation rule of the risk*

Finally, the contribution to the global risk of each terminal place of an elementary process is simply expressed by the "product" of the mean occurrence frequency of the corresponding accident by the associated damage. The first is the result of a simulation as above described and the second one is a quotation supported by an attribute of the states corresponding to an accident situation. With a simulation tool such as CPN Tools, the use of "monitors" allows us to memorize the time instant where the token enters the places associated with an accident situation and make a statistic to obtain an assessment of the occurrence frequency. The severity (or gravity) of the damage is an attribute of the accident situation places whose value is defined in a preliminary study. Let us recall that the damage may have several components whose measure units may be of very different types. In the case of domino effect, the resulting risk will be the sum of elementary contributions of each place associated with an elementary damage.

To assess the accident frequency of the whole process, as well as for the first elementary process (the frequency of the second process is equal to the frequency for the whole process), Monte-Carlo simulation with 100,000 stories has been carried out. The following data are considered:

– The exposition event of the first elementary accidental process occurs randomly once a year on average with a variance equal to 2,000 h. It stays valid for 1 week on average with a variance of 24 h.

– The initiating event of the first elementary accidental process occurs randomly twice a year on average with a variance equal to 2,000 h. It stays valid for 4 days on average with a variance of 24 h.

– The exposition event of the second elementary accidental process occurs randomly four times a year with a variance equal to 1,500 h. It stays valid for 10 days on average with a variance of 48 h.

The results obtained by Monte-Carlo simulation are presented in Table 6.1.

For one accident of the whole process, the first elementary process occurs on average 9.05 times by history. The minimum number is 1 (meaning that the domino effect to the whole accident process occurs at the first occurrence of the first accidental process) and the maximum number is 383. The accident of the whole process occurs only one time by story (average, minimum and maximum values equal to one and standard deviation equals to zero), because its occurrence is used as the condition to stop simulation.

The mean time to occurrence of the first elementary process is equal to $62.9 \cdot 10^5$ h with a half-length of 95% confidence interval equal to 2,257 h. The average frequency of the first accidental process occurrence is equal to $1.59 \cdot 10^{-5}$ by hour (once every 7 years).

The mean time to occurrence of the whole accidental process is equal to $1.35 \cdot 10^6$ h with a half-length of 95% confidence interval equal to 31,320 h. All statistical data from the minimum, maximum and mean values for the time to occurrence are identical because only one occurrence of this event takes place by story. The average frequency of the whole accidental process occurrence is equal to $7.38 \cdot 10^{-7}$ by hour (once every 154 years). This result can lead to think that the frequency of the domino effect may be neglected, however, it would not be forget that its minimal time occurrence is 8640 h (1 year approximatively). In a safety context, perhaps this last value must be considered instead of only average value.

	Statistics				
	Average	95% confidence interval half length	Standard deviation	Minimum value	Maximum value
	First elementary accidental process				
Number of occurrences by story	9.048	0.147	23.559542	1	383
Minimum time to occurrence [h]	17876.56	1685.078	268583.073	8538	36084260
Maximum time to occurrence [h]	820086.09	18586.397	2962467.164	8640	61565698
Mean time to occurrence [h]	62926.98	2256.962	359735.005	8640	36084260
	Accident of the whole process				
Number of occurrences by story	1	0	0	1	1
Minimum time to occurrence [h]	1354145.71	31320.003	4992063.862	8640	95178493
Maximum time to occurrence [h]	1354145.71	31320.003	4992063.862	8640	95178493
Mean time to occurrence [h]	1354145.71	31320.003	4992063.862	8640	95178493

Table 6.1. *Simulation results of accidental process*

6.6. Conclusion

The goal of the presented example is not of being realistic but to show how a chaining of successive elementary processes may be modeled and simulated. Of course, the confidence intervals can be improved by increasing the number of simulated stories, without great consequence on the whole simulation time that remains in the order of a few seconds. It was only an abstract example; now, in Chapter 7 we will show how to model a real accidental process of which the simulation could be made without difficulty with the generic models presented in the above section.

7

Illustrative Example

Let us take an illustrating example taken from the railway transport and more precisely safety aspects linked to the introduction of automation. Such devices are the concern of functional safety and their development must satisfy the relevant standard [IEC 02a, IEC 02b, IEC 07]. A preliminary risk analysis must be beforehand performed to assess the required SIL level the function must satisfy. The use of the method presented above can give an effective help to achieve such assessment.

7.1. Functional description

Figure 7.1 depicts the protection system by platform screen doors (PSDs) in a metro station. When the train enters the station, it must stop exactly in a position such that the doors of the train are exactly facing the doors of the platform screen. A set of devices (hardware and software) is in charge of this automation that must give a high degree of confidence to avoid any risk of injury for the metro users. Each station is provided with an input light signal automatically set to red at the slightest failure or doubt by the safety control system. An output light signal allows the train to leave the station.

Three types of train are riding on the track: a talkative train (with communication) without a driver, a talkative train with a driver (he has a reduced function: go/stop, doors opening, etc. he can be considered as a safety barrier) and finally a non-talkative train (without communication) with a driver of course. Talkative trains are designed according to the communication-based train control (CBTC) standard [IEE 04].

Figure 7.1. *The system train – screen doors*

This system of advanced technologies is under the control of electronic devices interconnected via wired or wireless communication networks. These devices include sensors, actuators, communication interfaces and signaling, a hierarchical system of wired electronics, programmable controllers and computers in charge of functional and safety actions.

It is not our purpose here to present either the whole description of such a complex system or its extended risk analysis, this could extend over too many pages. Among the most feared accident, we can mention the fall of a passenger on the track as a result of an unexpected opening of the platform screen doors and the injury of a passenger by an untimely door closing during the train stop at a station. We choose this last feared accident to illustrate the above presented modeling method.

7.2. Building of an accidental process

7.2.1. *First elementary process*

To study the risk of passengers being hurt by an untimely train door closing, we have to first consider two primary elementary entities, the vulnerable target (passengers) and the hazard source (doors). Figure 7.2 represents the process.

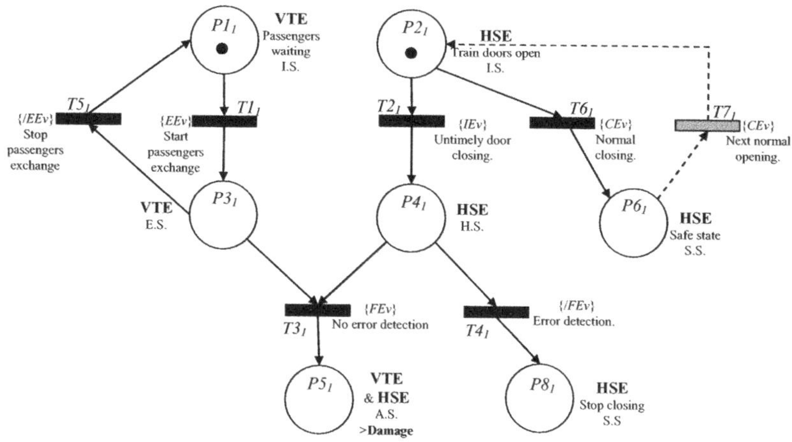

Figure 7.2. *Elementary process: "passenger hurt by untimely door closing"*

The generic model of an accidental process is recognized in the sets of places and transitions $\{P1_1, P2_1, P3_1, P4_1, P5_1\}$ and $\{T1_1, T2_1, T3_1\}$, the other elements being complements. In the initial situation (I.S.) of this process, the train is stopped, the doors have just opened ($P2_1$) and passengers are waiting on platform or in train ($P1_1$). Firing transition $T1_1$ means that a passenger exchange begins between the train and the platform. Passengers are consequently in exposition situation (I.S. in $P3_1$). In this situation, two events may occur. In normal conditions, all passengers enter or exit the train ($T5_1$) and the initial situation is reached ($P1_1$). In hazard conditions, $P4_1$ is marked because the initiating event "untimely door closing" associated with $T2_1$ occurred. Then, transition $T3_1$ is fired and the accidental situation (A.S.) is reached. Note that a feared event is associated with this last transition, it is the non-detection of the untimely door closing. This means that a detection mechanism (a technological device or at least the driver action) can avoid the accident (safety barrier). The token is then pushed in state $P8_1$ by firing transition $T4_1$. This state is safe (SS) for the target of this process: the passengers. Furthermore, it may be useful to model the normal behavior of the doors. When no untimely closing (initiating event) occurs before the normal closing order ($T6_1$) of the doors, the system enters a safe state $P6_1$ where the train starts to go and stops on the next station. We might be tempted to introduce a supplementary transition ($T7_1$), for example a control order, to push the token again in the initial situation and restart the accidental process. However, this will only be legitimate if in the following stations, the exposition event of passengers has exactly the same occurrence characteristic. It is also possible to consider that the transition $T7_1$ is timed to represent the mean time for the train to enter this same station. This is a question of refinement level of the description that will not be of course without effect on the simulation duration.

Let us now make some remarks about the characterization of the different events of this process and first the exposition ($T1_1$) and the end of exposition ($T5_1$) events. It is evident that these events are dependent on many factors characterizing the passenger traffic (attendance of the station, the day, the time, etc.). Statistical data given by the railway operator can help to characterize this event. Second, the events attached to $T3_1$ and $T4_1$ characterize the efficiency of the detection system. Such systems are often considered as characterized by a predefined probability of detection and of course the complementary probability on non-detection. If the detection time is neglected, the transitions become immediate and the assignment of the complementary probabilities to the transitions avoids the conflict problem.

The normal closing order ($T6_1$) occurs generally after a predefined time which constrains the occurrence of the "stop passenger exchange" event. For the simulation, this dependency must be considered in the definition of the event generator PN.

Finally, we suppose that the occurrence of the initiating event "untimely door closing" must be described in more detail as itself being the result of another accidental process.

7.2.2. *Second elementary process*

This process will, therefore, describe (Figure 7.3) how the event "untimely door closing" occurs. It is relative to the couple target–source, respectively, constituted by the train door and its control system (DCS). Places $P1_2$ and $P2_2$ are the initial situations: the train is stopped at station its doors are closed and the DCS is in operating state. An opening order ($T1_2$) put the doors in an exposition situation ($P3_2$) and a failure ($T2_2$) put the DCS in fail state ($P4_2$). As soon as $P3_2$ and $P4_2$ are marked, the transition $T3_2$ is fired generating the event "untimely door closing". In addition to producing this event, the PN memorizes the situation as accidental and a damage must be associated. Here, the damage will not only include the repair costs but also the consequences of the evacuation obligation of passengers and train (reduction of the track traffic and unavailability of the train).

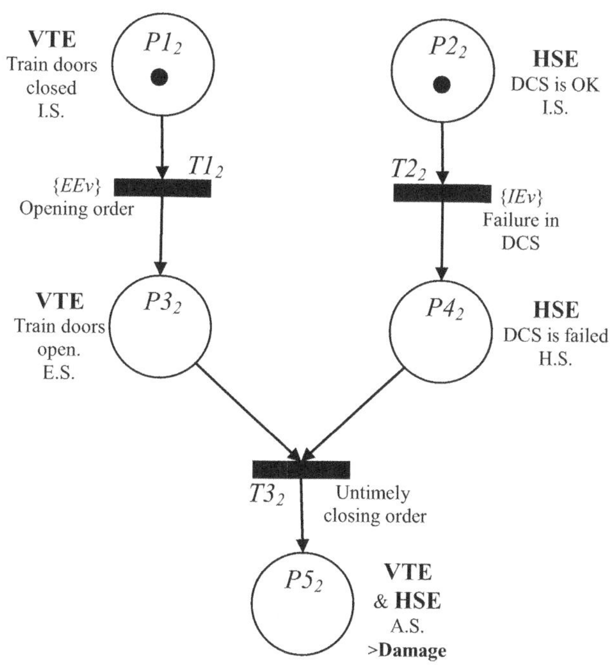

Figure 7.3. *Elementary process "untimely door closing"*

7.2.3. *Parallel process*

We saw that railway operators introduced screen doors on the platforms to protect passengers from hazards of train traffic. However, as well as train doors, they can be subject of failures, for example an untimely closing. In such a case, the possible accident may be described by another PN, a replication of this shown in Figure 7.2. We have, however, to remark that the target and the resulting damage are exactly the same so that both PNs could share the final place $P5_1$. This is a possibility but it would be a shame not to enjoy the modeling power of colored PNs. It would be then preferable to consider the subnet $\{P2_1, T2_1, T6_1, P4_1, P6_1, T3_1, T4_1, P8_1, \}$ able to represent both hazard sources: screen and train doors. Two token colors evolve in this part: "train" and "screen". The different transitions are sensitive to both colors but different events are to be considered according to the considered token color. Two synchronization events are then associated with transition $T2_1$, one of them concerning the color "train", the other one concerning the color "screen". These events are emitted by two separated instances of the PN presented in Figure 7.3.

It is important to note that in these two instances, the hazard sources are control systems, one is on board, the second on the platform. They appear as physically independent so that their failure occurrence can be characterized separately. However, the control architecture of the whole system shows that the door control systems are part of a higher level system, the automatic train control (ATC) in charge of the coordination of the various elementary functions. These doors control systems, in addition to possibly failing themselves, may also be considered as a hazard target exposed to the consequences of failures in the ATC. It will be then possible to further extend the modeling if the ATC system is included in the perimeter of the study.

7.2.4. *The whole model*

Figure 7.4 represents the whole model of accidental process. The first sub-PN (left top) represents the process leading to the untimely train doors closing, the sub-PN (right top) represents the untimely screen doors closing and the third one (down) represents the accidental process where passengers are hurt by one of the two doors type. The interaction of these three PNs is represented by the dashed arrows symbolizing the events emitted by transitions $T3_2$ and $T3_3$ and received by $T2_1$. A token of "train" color will fire $T2_1$ only on the occurrence of an event emitted by $T3_2$ and a token of "screen" color will fire $T2_1$ only on the occurrence of an event emitted by $T3_3$.

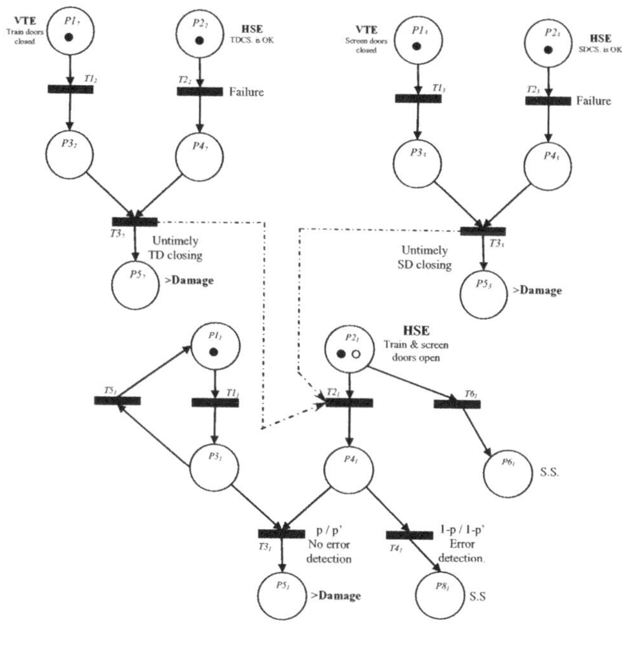

Figure 7.4. *The whole accidental process*

With regard to the simulation, a problem is the definition of the stop condition of a story. Some "dump" places (without output transitions) are present in the PN, this is generally a condition of non-liveness that will have a blocking situation as a consequence. It is usually necessary to check for this situation. However, as we can see the marking of $P5_2$ or $P5_3$ will block the whole process only if there is no passenger exchange when the untimely closing occurs. In the contrary case, the accidental process continues until the marking of $P5_1$. A story will then stop if a terminal place is reached except for $P5_2$ or $P5_3$ when $P1_1$ is marked.

7.3. Conclusion

In this short example of accidental process, we showed how it may be interesting to extend the study. This is one of the main advantages of PNs of being easily completed to put processes into a chain, to insert conditions on transition firings, etc. The implementation of such studies may be easily made with tools such as CPN Tools as we showed in chapter (section 6.5). The PN modules that were presented will easily be reused to perform the study of the whole set of accident scenarios and emphasize their interdependences. The instantiation mechanism allows us to describe all the identical processes by the same model and reduce then the risk of error in the replication and of a basic model.

8

Design and Safety Assessment Cycle

We presented in the previous chapters a method for a formal modeling of accidental processes whose main objective is to get an assessment of the mean occurrence frequency of accident situations. If the severity of the corresponding damages has been previously estimated, the risk assessment can then be accessed. These activities are essential in the frame of the risk management and more precisely in the design phase. In this chapter, we will only present in which steps of the whole design and safety assessment cycle our proposition takes place and will therefore emphasize the contribution of this ontological approach.

8.1. Five essential steps

In [MAZ 08, MAZ 07], we proposed decomposing the design and safety assessment process of a system in few main primary and complementary phases in accordance with the European directive [DIR 96]. Figure 8.1 gives a representation of the design and safety assessment cycle in the design activity of a safety-related system, all this being encompassed in the system engineering activity.

In this cycle, we are interested in the following five steps:

1) *Hazard study*:

This step is not intended to process the materialization of accident scenarios in details, but rather to highlight the various possible problems encountered during the lifecycle of the studied system. For this study, the analyst is guided by typical potential accident trees or checklists specific to the concerned study field mainly based on the learning feedback.

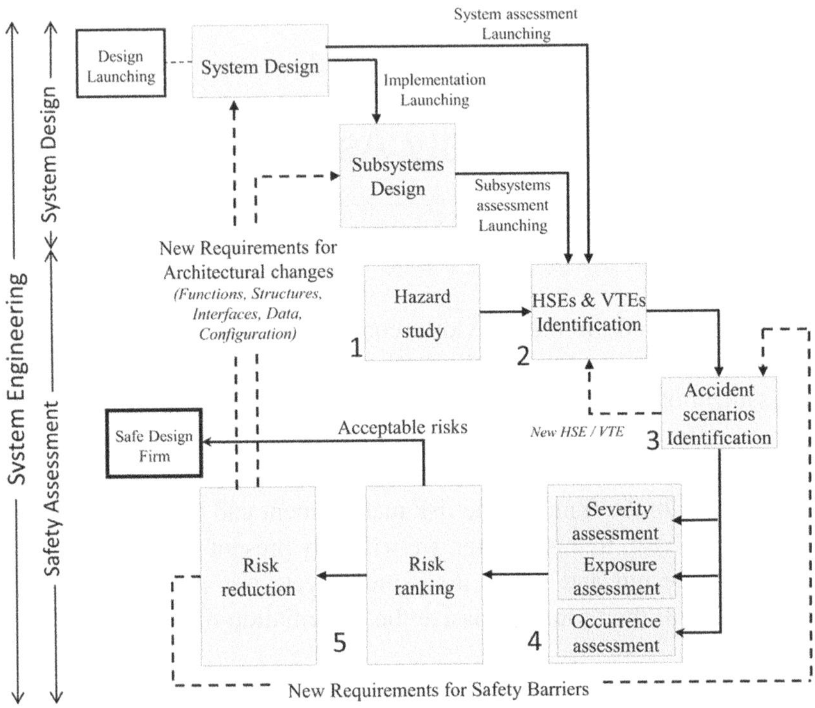

Figure 8.1. *Design and Safety Assessment Cycle*

2) *Deductive identification of HSE and VTE*:

For any entity of the system, the question of its identification to an HSE or a VTE must be asked.

To give an exhaustive character of this step, an entity-based hierarchical systemic classification of the global system must be established. It takes into account the socio-technical system (technological and human entities) and the environment (human, technological and natural entities), indicating internal interactions between these entities.

For each of the selected entities, the research of the potential initiating and exposition events must be identified.

3) *Inductive identification of the accident scenarios*:

From each hazard source entity (HSE) identified through the previous phase, the analyst tries to develop the different ways inducing potentially a hazardous event and deduces the corresponding hazard spaces. Thus, for each hazard space, it is necessary to identify the vulnerable target entity (VTE) whose vulnerability space may be concerned. For each couple HSE/VTE thus selected, the feared event must be defined and the accident situation must be described. A systematic research of possible interactions between elementary processes must be conducted to identify all the VTEs submitted to a same HSE, all the HSEs acting on a same VTE and all the event identities responsible for elementary processes sequencing (propagation phenomena).

4) *Risk assessment*:

The risk assessment covers four complementary steps: determination of the possible consequences, assessment of the severity, assessment of the occurrence frequency of the feared event and calculation of the corresponding risk as a combination of exposure, occurrence and severity. In traditional approaches, the assessment of these three parameters is made by raking (not without subjectivity) among some predefined levels, for example: severity $\in \{S1, S2, S3, S4\}$, occurrence $\in \{O1, O2, O3, O4\}$ and exposure $\in \{E1, E2, E3\}$. This presentation does not consider the interaction between elementary processes, for example a propagation between entities at different successive hierarchical levels (for example, domino effect). In such cases, all damages of the successive VTE are cumulated but each of them is not affected of the same occurrence frequency and the frequency of the ultimate accident is dependent on all the previous ones. This interaction is very rarely analyzed.

5) *Risk reduction*:

It is a risk-covering phase, while engaging the appropriate actions to avoid the causes, mitigate the effects or limit the consequences. In addition, it is suitable to reduce the exposure factor. Indeed, it is required to build the following defense-in-depth lines: preventive measures (to eliminate HSEs or to reduce $IEvs$ occurrences), protective measures (to mitigate the impacts on $VTEs$) and recovering measures (to go back faster to the initial situation).

The risk assessment approach presented in Chapter 6 does not recover the whole cycle. It essentially takes place in phases 2, 3 and 4. Phase 2 is totally covered and beyond the identification of phase 3, a formal model of the scenarios is proposed. In phase 4, the assessment of a risk frequency occurrence is the result of a simulation based on the formal model defined in the previous phase. In phase 5, the model of new entities introduced as reaction means preliminary evaluated risks are defined and introduced in the next iteration of phase 4.

Although it does not recover the whole cycle of the design and safety assessment cycle, this approach contributes to essential phases whose objective is the risk assessment and it is precisely on this activity that decisions to be taken will be based (mitigations, protections, barriers or eventually modification of the initial architecture of the system). It then represents a significant and safe contribution.

8.2. Ontological interest

Previously, we positioned the proposed approach in the design and safety assessment cycle of a system. In Chapter 5, we highlighted the alleged ontological character of this approach. From the experience illustrated by the example in the railway domain (Chapter 7), we can conclude on the compliance to the criteria defined by Grüber (see Introduction to Part 2) for a good ontology.

For clarity and coherence purposes, the ontology concepts are presented in a deductive way. All the concepts introduced are regrouped as per their semantic relationship. In [MAZ 08], each concept is presented following specific rules:

– a presentation of the different definitions starting with the literal meaning (dictionaries, etc.), then the definitions given by system engineering and safety experts, then the definitions given by specialized work-groups [AQS 96, SEM 97, PRI 03], then the definitions proposed by the French national standards (NF) or British [GEN 92], European (EN, etc.) and international [MAN 02] and finally, if necessary, the definitions issued from legislation (EU directives, law, decree, etc.);

– a summary to detect the differences and similarities of the reported definitions and sometimes review of some inconsistencies or contradictions;

– a proposition of a new definition if none of the reported definitions should be fully adopted.

Another criterion for ontology is the extendibility, in other words, the enrichment with new concepts for special uses based on the existing ones, without any revision of the existing definitions. This has been demonstrated, for example, about the introduction of a safety barrier in the elementary accidental process or the accumulation model of the exposition to a danger due to the addition of tokens in a place associated with an exposition situation.

Finally, with regard to minimal encoding bias and minimal ontological commitment purposes, the genericity of the models has been illustrated in [MAZ 08] with its application in several fields (railway, road, manufacturing, financial, epidemiological, juridical, etc.), without additional instantiation commitment or encoding bias on the basic concepts.

Conclusion to Part 2

We presented an example of risk analysis based on the modeling of elementary processes interacting together. Additionally to an improved clarity in the phenomena description, PNs bring the advantage of a formal model whose properties would be exploited. We choose to direct the obtained model to simulation for a quantitative assessment of the risk. One of its main interests is the ability to be also extended for the effectiveness assessment of a safety barrier or of the evaluation of its cost/effectiveness ratio.

With PNs being commonly used for the modeling of automated systems behavior as well for design as for performance analysis, coupling the risk analysis with the functional model of a system becomes easy. We can study, for example, the effect of the constraint imposed by the implementation of a safety barrier on the functional performances. From these viewpoints, it is a useful decision aid tool in the establishment of safety systemic strategies in an industrial plant or company. In addition, the use of PN models can also open a door toward the quantitative assessment tools of dependability where they become to be commonly used.

Despite the progress made by the introduction of formal models in risk analysis, the approach remains dependent on the limits inherent to its input data: event occurrence probabilities and consequential damages. The improvement of the first ones would bring risk assessment closer to reliability assessment provided that realistic stochastic distribution laws would characterize the various events of the accidental processes. Although we were able to introduce stochastic delays in CPN Tools models, there has been no question of stochastic Petri nets, models commonly used for reliability, maintainability or availability quantitative assessment. These models will be described more specifically in the following sections.

PART 3

Stochastic Petri Nets

Introduction to Part 3

In Parts 1 and 2, we introduced Petri nets and gave application examples based on timed and synchronized PNs. Although we presented timed and synchronized models for simulations using random event generators, we did not explicitly talk about stochastic variables. Stochastic variables in dependability assessment are time variables, for example time to fail or time to repair, of an entity as defined in all books dealing with dependability and as recalled in the first book of this series [AUB 15]. It is then natural to think of timed PNs as the most suitable model for stochastic extension. Despite the fact that stochastic transition-timed PNs exist, it is not the model most commonly used to describe systems including stochastic time durations. The determinant element to represent the behavior of such systems is the firing of a transition that is to be considered as a random event whose occurrence is characterized by a rate, elementary probability for the transition to be fired. These are stochastic Petri nets.

9

Basic Concept

9.1. Introductory example

Let us consider the example of Figure 1.5 and reconsider the need to introduce the time in the model, especially when the duration of operations associated with transitions is of a stochastic nature. In this example, let us consider that transition T_1 corresponds to a machining operation on parts stored in an input stock represented by place P_1. In this example, reproduced in Figure 9.1, two machines are available for the operation (two tokens in the initial marking of P_3). Each machine can operate on a part if the input stock is sufficient but each machine can fail (transition T_2) and can be repaired. The duration of the machining is not at all constant but depends on a lot of physical parameters, such as quality of parts, wear of tools, lubrication of actuators, etc. It would be possible to associate a firing delay to this transition as we did for timed PNs, in addition to a stochastic character. The obtained model is known as the stochastic timed PN and involves the notion of reserved marking [DAV 89, DAV 92].

The stochastic PN, which is usually preferred, associates with each transition of an autonomous PN a firing rate α such that $\alpha.dt$ is the probability for the transition to be fired between the time instants t and $t + dt$ knowing that the transition is enabled at t.

In our example, α is the machining rate, λ is the failure rate of a machine (the two machines are supposed as identical) and μ is its repair rate. In the case of α, λ and μ are constant (exponential distributions of the machining, operating and repair times). These rates are respectively equal to the inverse of the mean machining time, the inverse of the mean time to failure (MTTF) and the inverse of the mean time to repair (MTTR).

If the PN of Figure 9.1 was an ordinary PN, there would be a conflict between transitions T_1 and T_2. Here, since these transitions are competing, the fired transition will be the one whose associated event occurs first (end of the machining or failure of a machine).

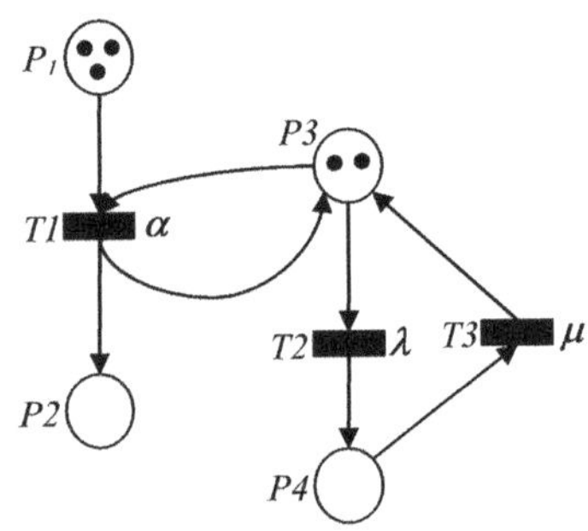

Figure 9.1. *Stochastic Petri net of the machining system with two machines*

9.2. Formal definition

DEFINITION 9.1.– A stochastic Petri net (SPN) is a couple $R_S = \langle R, \Lambda \rangle$ such that:

– $R = \langle P, T, w^-, w^+, M_0 \rangle$ is a generalized PN;

– $\Lambda : T \to \mathbb{R}^+$ is the application associating with each transition a firing rate: $\Lambda(T_j) = \lambda_j$.

Notion of n-enabled transition

In a generalized PN, a transition T_j is n-enabled if:

$$n \leq min(M(P_i)/w^-(P_i, T_j) < n + 1, \qquad P_i \in I(T_j)$$

Remembering that $M(P_i)$ is the marking of P_i, $I(T_j)$ the set of input places of T_j and $w^-(P_i, T_j)$ the weight of the arc (P_i, T_j).

In the case where all the arcs are 1-weighted, the condition becomes:

$$min(M(P_i)) = n, \qquad Pi \in I(Tj).$$

In our example, the transition T_1 is 2-enabled because $I(T_1) = \{P_1, P_3\}$, $M(P_1) = 3$ and $M(P_3) = 2$. Transition T_2 is also 2-enabled because $I(T_2) = \{P_3\}$ and $M(P_3) = 2$.

Clearly, when n entities compete together for the execution of a same task, the mean execution time of this task is divided by n, meaning that the firing rate of the transition is multiplied by n. In our example, the failure rate of one of the two entities is doubled just as the machining rate when two machines are available and at least two parts are in the input stock.

Evolution of the marking

All the transitions enabled by a marking M compete for firing, the transition fired corresponding to the associated event which will occur first.

To represent the operation of the system, for example during a simulation, the marking of a SPN will evolve in the following way: for a given marking, many transitions compete to be fired. For each of them, a stochastic draw of the firing duration $d_j = 1/\alpha_j$ is performed. For an n-enabled transition, if the firing rate of a transition T_j is α_j, then the draw is made with a rate $n.\alpha_j$. After a time equal to the shorter of these durations, the corresponding transition is fired leading to a new marking and a new set of enabled transitions on which a new draw is made, and so on.

10

Semantics, Properties and Evolution Rules of an SPN

We will first recall some usual formulas before demonstrating. Let:

– M be an accessible marking for the autonomous bounded PN underlying an SPN;

– $\alpha_j(M)$ be the firing rate associated with the transition T_j for the marking M (if Tj is n-enabled for M, then the firing rate is $\alpha_j(M) = n\alpha_j$).

The time duration D_j in the state M between the time instants where T_j is enabled and then fired, has a probability distribution and a density functions:

$$H_j(t) = 1 - e^{-\alpha_j(M).t} \qquad h_j(t) = \alpha_j.e^{-\alpha_j(M).t}$$

Let $T(M)$ be the set of transitions enabled by the marking M.

If $T_k \in T(M)$, then the firing conditional probability of this transition from the marking M is:

$$\Pr(T_k/M) = \frac{\alpha_k(M)}{\sum \alpha_j(M)}, j|T_j \in T(M) \qquad [10.1]$$

and the sojourn time in the state M follows an exponential law of parameter:

$$\alpha(M) = \sum_j \alpha_j(M), j|T_j \in T(M) \qquad [10.2]$$

This gives a mean sojourn delay in the marking M:

$$\overline{D}_j = \frac{1}{\alpha(M)} \qquad [10.3]$$

10.1. Conservatism properties

The marking $M(t)$ of an SPN is a random vector of dimension p whose component i is the random variable representing the marking of the place P_i.

Let $N_j(t)$ be the random number of the T_j transition firings in the time interval $[0, t[$. The vector $N(t)$ whose $N_j(t)$ is the jth component is the firing random vector.

At any time, the marking and firing processes are interlinked by the relation (see fundamental equation [1.4] of PNs, where W is the incidence matrix): $M(t) = M(0) + W \cdot N(t)$, giving the mathematical expectation:

$$E\{M(t)\} = M(0) + W \cdot E\{N(t)\}$$

10.1.1. *Conservatism of the mean marking in steady state*

Let X be a P-semi-flow of the underlying PN (recall: $X^T \cdot W = 0$), then with analogy of autonomous PN , $E\{M(t)\}$ satisfy the relation:

$$X^T \cdot E\{M(t)\} = X^T \cdot M(0) \qquad \forall t$$

Let us suppose the marking process to be ergodic and stationary (it converges to the same finite limit $\overline{M}$ in temporal and probabilistic averages), then the vector $\overline{M}$ is the mean marking vector in steady state.

It satisfies the property $X^T \cdot \overline{M} = X^T \cdot M(0)$ (mean marking invariant) meaning that the weighted sum of mean markings in a conservative component equals the same weighted sum of the initial marking of the places of this conservative component.

10.1.2. *Conservatism of the flow in steady state*

The firing process being supposed as ergodic, we will denote $\overline{F}$ the vector which is the finite limit in temporal and probabilistic average of the firing random vector $N(t)$.

$\overline{F} = \left[\overline{f}(T_1), \overline{f}(T_2), ..., \overline{f}(T_m),\right]^T$ is known as the mean frequency vector of transition firing in steady state.

This vector satisfies the following relation:

$$W \cdot \overline{F} = 0$$

F^* is a transition firing invariant (T-semi-flow).

This relation expresses the conservatism of the token flow in an SPN.

As $W = W^+ - W^-, \quad W^+ \cdot \overline{F} = W^- \cdot \overline{F}$

This relation states that for any place of a stochastic PN, the incoming flow of a place equals the outgoing flow of the same place.

10.2. Mean sojourn time in a place of a SPN

With the same hypothesis on the marking and firing processes, the Little formula [10.4] gives the mean sojourn delay of a token in the place P_i, denoted by $\overline{D}(P_i)$:

$$\overline{D}(P_i) = \frac{\overline{M}(P_i)}{w_i^+ \cdot \overline{F}} \qquad [10.4]$$

where w_i^+ is the ith line of W^+.

The mean sojourn time equals the ratio of the mean marking of this place in steady state over the sum of the mean firing frequencies of the transitions upstream this place, each of them being multiplied by the weight of the corresponding arc.

As we will see later, the calculus of this time requires the resolution of the associated Markov process.

10.3. Equivalent Markov process

When the underlying PN is bounded, the reachability graph (n distinct values of the marking vector) is homogeneous to a continuous time Markov process of size n, which is obtained by associating with each arc a rate depending on the fired transition rate and of the marking of the input places of this transition (n-enabled transition). Let $\mathbb{P}(t) = [\Pr_1(t), \Pr_2(t), ..., \Pr_n(t)]$ be the probability line vector (size n) where the ith component $\Pr_i(t) = \Pr(M_i)$ is the probability of being in the marking M_i (state i).

The Chapman–Kolmogorov (C–K) vectorial equation of this process is:

$$\left[\frac{d\Pr_1(t)}{dt}, \frac{d\Pr_2(t)}{dt}, ..., \frac{d\Pr_n(t)}{dt}\right] = \left[\Pr_1(t), \Pr_2(t), ..., \Pr_n(t)\right] \cdot \mathbb{A} \qquad [10.5]$$

usually denoted as:

$$\dot{\mathbb{P}}(t) = \mathbb{P}(t) \cdot \mathbb{A} \qquad [10.6]$$

where:

$$\left[\Pr_1(t) + \Pr_2(t) + ... + \Pr_n(t)\right] = 1 \qquad [10.7]$$

$\mathbb{A}$ is the infinitesimal generator matrix of the system (the term a_{ij} of $\mathbb{A}$ is the transition rate from the state (marking) i to the state (marking) j). Solving this differential equation system requires the knowledge of the probability vector $[\Pr_1(0), \Pr_2(0), ..., \Pr_n(0)]$ at the initial instant time $t = 0$. This solution is given by the following relation:

$$\mathbb{P}(t) = \mathbb{P}(0) \cdot e^{\mathbb{A}.t} \quad [10.8]$$

If the continuous time Markov chain isomorphic to the SPN is ergodic, the asymptotic behavior of the system tends to a single limit distribution independent of the initial conditions:

$$\Pi = \mathbb{P}(\infty) = \lim_{t \to \infty} \mathbb{P}(t). \quad [10.9]$$

The vector Π, state probability distribution in steady state, is known as the stationary distribution of probabilities.

In steady state, the fundamental equation becomes: $0 = \Pi{\cdot}\mathbb{A}$, and the stationary distribution Π may be obtained accounting that: $\Pi{\cdot}\mathbb{I} = 1$ where $\mathbb{I}$ is a summing vector with all components equal to 1.

The components of the vector $\Pi = [\pi_1, \pi_2, ..., \pi_n]$ are the stationary probabilities of being in the different states of the Markov chain, that is to say the different markings of the SPN.

Knowing the mean probabilities of the SPN markings given by the vector Π, many performance indicators may be deduced:

– the mean markings of the places $P_i \in P$

$$\overline{M}(P_i) = \sum_{k=1}^{n} M_k(P_i) \cdot \pi_k \quad [10.10]$$

– the mean firing frequencies of the transitions $T_i \in T$

$$\overline{f}(T_j) = \sum_{k=1}^{n} \lambda_j(k) \cdot \pi_k \quad [10.11]$$

k such that T_j is fireable from M_k, $\lambda_j(k)$ depending on M_k;

– the mean sojourn delays of the token in the places in steady state given by the Little formula [10.4].

10.4. Example of SPN for systems dependability modeling and assessment

A hydraulic system is composed of two identical pumps operating in parallel (redundancy) and of a valve controlling the global flow (Figure 10.1). Let us denote, respectively, λ_p, μ_p and λ_v, μ_v, the failure and repair rates of the pumps and of the valve. It is supposed that the pipes are failure free.

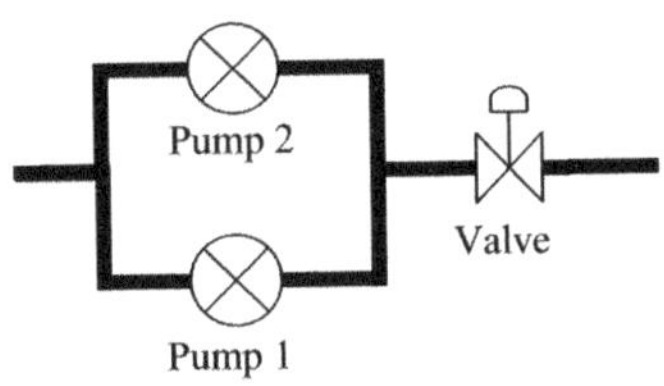

Figure 10.1. *The hydraulic system*

The SPN modeling the dependability of the system is composed of two disconnected parts each of them representing on the one hand the behavior of the pump set and on the other hand the behavior of the valve (Figure 10.2). On the right loop, two places P_1 and P_2 represent, respectively, the pump number in operating state and the pump number in failed state. Transitions T_1 and T_2 represent, respectively, the failure of a pump and the repair of a pump (they are undifferentiated).

In the left loop, P_3 and P_4 symbolize the state of the valve, operating or blocked, the transitions T_3 and T_4 representing its failure and repair events. It is supposed that the repair of any of the three components is possible at any time. The initial marking shows that all elements are operating at the system start.

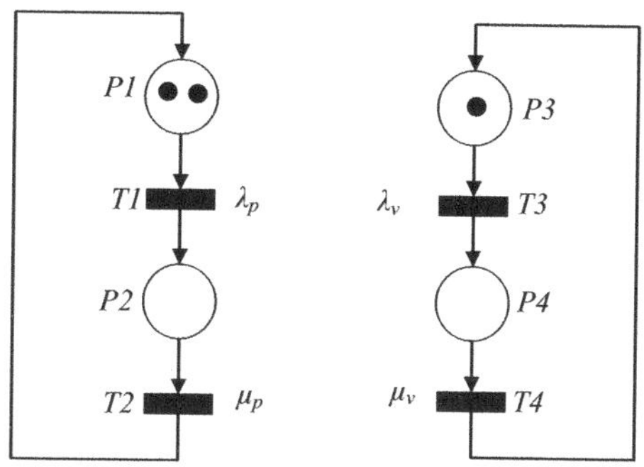

Figure 10.2. *PN of the hydraulic system*

The incidence matrix is:

$$W = w^+ - w^- = \begin{bmatrix} -1 & 1 & 0 & 0 \\ 1 & -1 & 0 & 0 \\ 0 & 0 & -1 & 1 \\ 0 & 0 & 1 & -1 \end{bmatrix}$$

There are two conservative components $\{P_1, P_2\}$ and $\{P_3, P_4\}$, the corresponding P-semi-flows are $M_1^T = [1, 1, 0, 0]$ and $M_2^T = [0, 0, 1, 1]$ and the mean marking invariants are $\overline{M}(P_1)+\overline{M}(P_2) = 2$ and $\overline{M}(P_3)+\overline{M}(P_4) = 1$.

There are two repetitive components $\{T1, T2\}$ and $\{T3, T4\}$, the T-semi-flows are: $F_1^T = [f_1, f_2, 0, 0]$ and $F_2^T = [0, 0, f_3, f_4]$ as each time T_1 is fired (respectively T_3) T_2 will be fired (respectively T_4) and the firing invariants give: $\overline{f}_1 = \overline{f}_2$ and $\overline{f}_3 = \overline{f}_4$.

In this example, it is easy to model a maintenance policy giving the priority to the element which failed first. The repairer is modeled by an additional place P_5, as shown by Figure 10.3, whose token represents its availability. This token will condition the firing of the transitions T_2 and T_4. No repair will be started as long as the repairer is engaged for another one. This is another example illustrating the ability of PN models to account for the need of memory due to the marking concept.

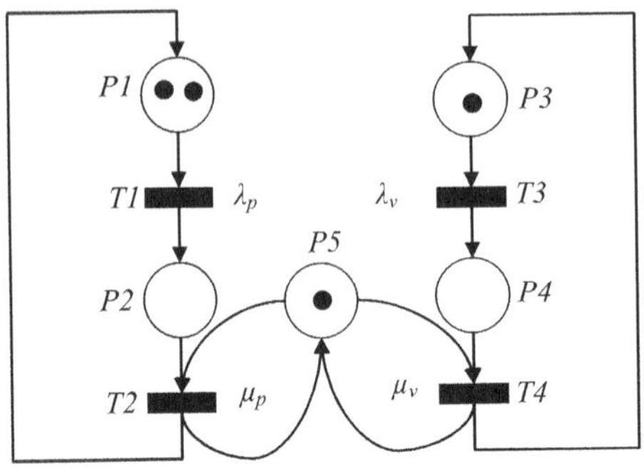

Figure 10.3. *PN with repairer sharing*

The graph of the accessible markings of the PN of Figure 10.2 is homogeneous to the Markov graph given by Figure 10.4.

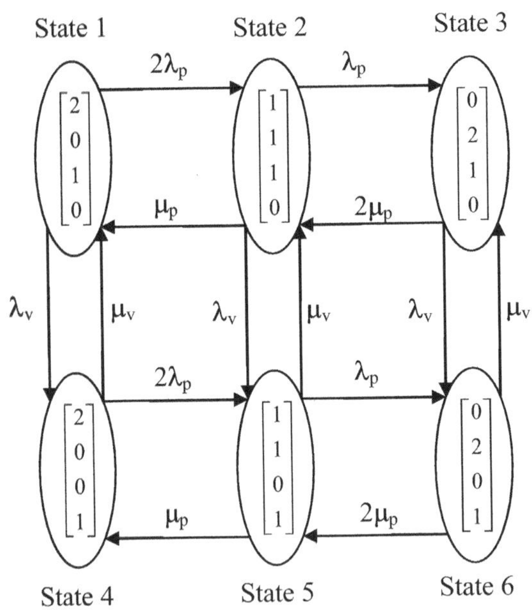

Figure 10.4. *The reachability graph homogeneous to a Markov graph*

The state transition matrix is:

$$\mathbb{A}=\begin{bmatrix} -(\lambda_v+2\lambda_p) & 2\lambda_p & 0 & \lambda_v & 0 & 0 \\ \mu_p & -(\lambda_p+\mu_p+\lambda_v) & \lambda_p & 0 & \lambda_v & 0 \\ 0 & 2\mu_p & -(2\mu_p+\lambda_v) & 0 & 0 & \lambda_v \\ \mu_v & 0 & 0 & -(\mu_v+2\lambda_p) & 2\lambda_p & 0 \\ 0 & \mu_v & 0 & \mu_p & -(\mu_v+\lambda_p+\mu_p) & \lambda_p \\ 0 & 0 & \mu_v & 0 & 2\mu_p & -(\mu_v+2\mu_p) \end{bmatrix}$$

With the following values: $\lambda_p = 10^{-2}; \mu_p = 0.1; \lambda_v = 10^{-3}; \mu_v = 0.2$ (not necessarily realistic), solving the equation system gives:

$\pi_1 = 0.82233; \pi_2 = 0.16447; \pi_3 = 8.2233 \times 10^{-3}; \pi_4 = 4.1117 \times 10^{-3}; \pi_5 = 8.2233 \times 10^{-4}; \pi_6 = 4.1117 \times 10^{-5}$

– $\pi_1 = 0.82233$ is the stationary probability for the normal operating mode;

– $\pi_2 = 0.16447$ is the stationary probability for the degraded operating mode;

– $\pi_3 + \pi_4 + \pi_5 + \pi_6 = 0.0132$ is the stationary probability for a non-functioning mode.

The mean firing frequencies, the mean markings and the mean sojourn times in places may be then calculated:

$$\overline{f}_1 = \overline{f}_2 = 2\lambda_p.\pi_1 + \lambda_p.\pi_2 + 2\lambda_p.\pi_4 + \lambda_p.\pi_5 = 1.8182.10^{-2}$$

$$\overline{f}_3 = \overline{f}_4 = \lambda_v.\pi_1 + \lambda_v.\pi_2 + \lambda_v.\pi_3 = 9.95.10^{-4}$$

$$\overline{M}(P_1) = 2.\pi_1 + 1.\pi_2 + 0.\pi_3 + 2.\pi_4 + 1.\pi_5 + 0.\pi_6 = 1.818$$

and so on: $\overline{M}(P_2) = 0.1818$ $\overline{M}(P_3) = 0.995$ $\overline{M}(P_4) = 4.975.10^{-3}$

$\overline{D}(P_2) = 10$ $\overline{D}(P_3) = 1000$ $\overline{D}(P_4) = 5$

11

Simplification of Complex Models

The approach presented in this chapter comes from the PhD research work of R. Schoenig [SCH 06, SCH 04] in the framework of an industrial cooperation.

11.1. Introduction

As we already discussed, Petri nets (PNs) may be considered as a powerful means for the modeling of systems with stochastic behaviors. Through all the assumptions required, the reachability graph can be made homogeneous to a Markov graph. When the complexity of the model increases, especially due to interactions in the system between functional and dysfunctional behaviors (one problem of the so-called dynamic reliability), the reachability graph can become huge in size. The question of how to go about its reduction may then be asked. Aggregation of states into subsets is a possible way. Known aggregation techniques of Markov graphs used to reduce the size of state spaces are relative to graphs whose transition rates belong to two separated sets of values. The aggregation process consists in splitting the Chapman–Kolmogorov equations of the whole problem into two subsets, the first subset accounting for a "slow" time scale and the second subset accounting for a "fast" time scale. Most of the methods require the construction and the analysis of the transition rate matrix. The existence of two time scales allows the introduction of a small parameter ε in the equations in which the generator matrix $\mathbb{A}$ can be written into the following form: $\mathbb{A}(\varepsilon) = \mathbb{A}_0 + \varepsilon \times \mathbb{A}_1$.

Three problems are inherent to this approach:

– the preconstruction of an exhaustive Markov graph is very difficult for systems issuing from industry because they rapidly generate a combinatorial expansion;

– the exclusive use of exponentially distributed events (constant rates) and the difficulty to take the events describing the logical behavior of the system into account limit the design power;

– the designer is not always accustomed to handling Markov graphs (this needs an *a priori* knowledge of the whole state set) and a representation of the system in a high-level formalism (such as PNs or more generally finite state automata) comes more naturally.

That is why we propose to combine both resolution techniques: simulation and Markov graphs. An aggregated Markov graph is built from considerations about the structure of the system. Indeed, the subsets of states are directly deduced from the knowledge of the operating modes (nominal, reconfiguration, etc.). Thus, the simulation is used only with regard to the fast dynamic of the system in order to determine the equivalent rates of the aggregated graph. Using simulation on a high-level of formalism avoids the introduction of Markovian hypothesis which is of no sense for the representation of determinist events. Thus, the manipulation of both techniques presents two advantages. On the one hand, it allows a reduction of simulation time and, on the other hand, it allows a reduction of the combinatorial expansion due to the "hybrid" aspect of these systems combining discrete and continuous variables.

11.2. System modeling

Introduced for the design and the control of discrete event systems, PNs are able to model hybrid and dynamic behaviors of a system as well as its control, its component failures and repairs, and the interactions between all these aspects. To account for such complexity, three models will be considered:

– controlled process modeling. This part is represented by a set of algebraic-differential equations reflecting the evolution of continuous physical variables (i.e. liquid level in a tank, car speed, etc.);

– control system modeling by control interpreted PNs. This state/transition formalism is appropriate to express the discrete feature of control systems. The control model fulfills numerical orders for actuators according to the measured information arising from the process. The interaction between the interpreted PNs and the equations of the process is formally defined by the activation of the functions associated with the places of the PN [JAM 01]. Let us note that this class of PNs is safe (one token per function);

– modeling of the physical element failure (sensors, actuators, calculators, etc.). A failure can alter the nominal evolution of the control PN by locking, premature firing transitions or erroneous firing transitions. Stochastic PNs can model the random occurrence of failures according to probabilistic laws.

Figure 11.1 shows this modeling. A Monte-Carlo simulation of the whole model (both discrete and continuous parts) reproduces the real operation of the process (featured by its own dynamic), and the control subsystem behavior according to both process state and current failure. The marking of the place Fail State (FS) in Figure 11.1 means that a feared event arises. A quantitative study allows assessing the occurrence probability by simulation.

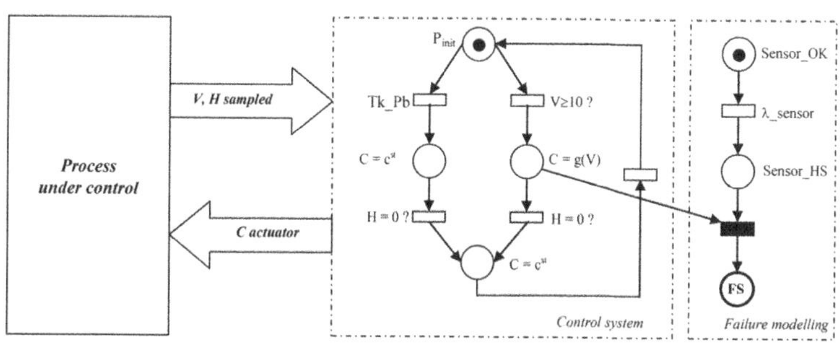

Figure 11.1. *Example of a control system modeling and its failures*

Nevertheless, even if theoretically this approach seems easy to be applied by virtue of its simplicity, its effectiveness remains a real problem due to the long simulation time required to obtain precise results. Indeed, to build up narrow confidence intervals for the results, a large number of simulator cycles are required. As for analytical methods based on state graphs, it is proving difficult to put them into practice for complex hybrid systems, because of the generated combinatorial expansion. Moreover, if we consider the set of possible values of physical variables, the number of states is theoretically infinite. A sampled model of the continuous part is then necessary.

11.3. Presentation of the quantitative analysis method

The proposed approach (evaluating the occurrence probability of identified feared events) consists of building up a reduced Markov graph. An exhaustive graph is, however, complex to generate and is also difficult to exploit because of its size (problems of storage, data processing, etc.). Instead of showing each elementary state of the system, the reduced graph depicts only macrostates relating from the union of elementary states sharing a feature, for example their membership to a same functional/dysfunctional mode (nominal operation, degradation, malfunctioning, etc.). This union of states provides a better overview and interpretation of the graph, because a limited number of macrostates are represented. Let us now present the theoretical principles to generate such a graph.

11.3.1. *Steps to obtain an aggregated Markov graph*

Let us now consider a control system formalized in PN. This system can be fitted with reconfiguration possibilities, providing a degraded operation after detecting one or several failures. The reachability graph is made up of a set of nominal states, degradation states and failure states. The transitions between two modes are fired on failure occurrence. We will make the following hypothesis regarding the modeling:

– to simplify the explanation of the methodology, we consider the system as not repairable. Thus, the states reached after feared events are absorbing;

– the system is supposed to be verified and validated first. In a given functional mode (nominal or degraded), the system has the appropriate properties in the absence of failure (liveness, determinism, no deadlock, etc.) (see [JAM 01]);

– when a failure occurs, the system switches either to another operating mode, or to a feared state.

It is easy to obtain a state graph from the reachability graph of the PN due to the net structure. As the PN of the control system has only one token, the reachability graph and the state graph are merged together.

11.3.1.1. *Generating the reachability graph*

In a didactic aim, the system is supposed to be simple enough so that it can be depicted by a reachability graph. Figure 11.3 shows an example of this.

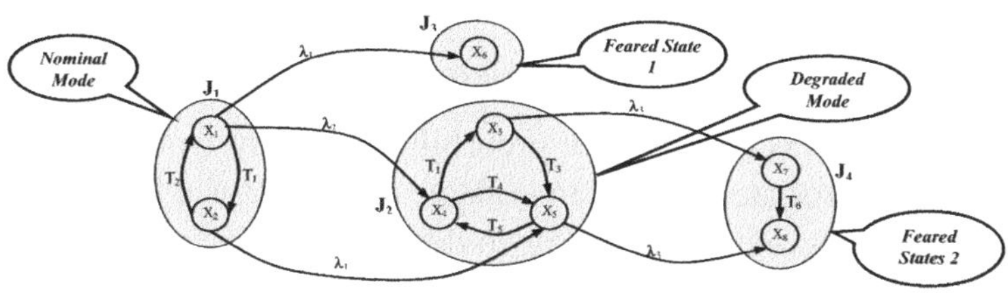

Figure 11.2. *Example of reachability graph*

The states x_i are lumped together according to their membership to a common operating mode. Therefore, four sets of states J_j are identified. These macrostates define a nominal mode, a degraded mode and two sets of malfunctioning mode.

The transitions T_i of the PN are fired when the place upstream is marked, and the condition linked to the transition is fulfilled (condition on a real value, threshold crossing, etc.). These transitions are included in the macrostates, because their firings correspond to a forecast evolution of the control system in an operation mode (defined by the designer), whether a nominal or degraded one.

The arcs identified by λ_i mean that a failure occurs, which makes the system switch to degradation state if a fault tolerance device has been provided, or to a feared state ("Feared State 1" or "Feared States 2"). Therefore, a fault occurrence makes the system change from one macrostate to another.

Let us remark the particular structure of the graph: after deleting the arcs λ_I, partial graphs are obtained which represent the possible evolutions of the system in a given mode (functioning or malfunctioning mode).

The sub-graphs related to the nominal and degraded modes are necessary and strongly connected by virtue of the underlying PN structure (the PN of the system is live, and for each operating mode, the relating subgraph has the appropriate properties). This remark is not valid for a subgraph relating to feared states.

11.3.1.2. *Construction of a complete Markov graph*

Here, the aim is to assess the probability of the system being in an absorptive macrostate, in order to evaluate the occurrence frequency of the feared events and, if necessary, to decide corrective actions on the system to decrease their occurrence probability. A fail-safe system can be designed step by step.

A Markov process with discrete state spaces and continuous time will be generated from the initial reachability graph expressing the whole state and the transitions between these states.

Earlier, to find safety indicators from the system modeling in PN, it was necessary to perform a co-simulation of the discrete part relating to the control system, and the continuous part relating to the controlled process (see [CHA 99] and [CHA 98]). Nevertheless, we want to establish an entirely analytical modeling based on a Markov process so as to overcome high simulation time due to the rarity of fault occurrence. Thus, graphical modeling can be translated into simultaneous differential equations, and the resolution will give the requisite probabilities.

An analytical resolution is not possible from the initial reachability graph of the underlying PN model, because of the conditions tied to the transitions. For this reason, an entirely probabilistic model will be generated which can then be easily translated into a differential equation (Chapman–Kolmogorov).

The approach consists of building up an analytical model, based on a Markov structure, whose stochastic variables are identical to the initial modeling. In order to build this Markov process, we will use the reachability graph of Figure 11.3.

The transitions T_i are replaced by transitions with an exponential distribution. The transition rates will be called α_i for each transition T_i. Only transitions included in transitional macrostates are considered. Hence, we obtain a graph similar to the one presented in Figure 11.3.

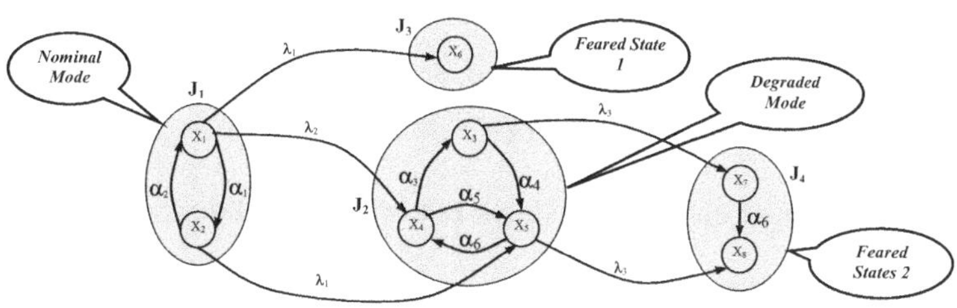

Figure 11.3. *Modeling into a Markov process*

The following technique will be used to calculate the transition rates α_i. First, the system is assumed to be in one of the macrostates (i.e. a given functional mode), and to stay in it (i.e. no failures); then, the steady-state probability in each elementary state x_i of the macrostate is considered to be identical between the initial graph and the new analytical graph which is wholly randomized with exponential transition rates. Each macrostate is investigated independently of each other.

If we suppose the transition rates to be known, two orders of magnitude can be distinguished: transition rates α_i and failure rates λ_i. The first ones depend on the sojourn time in the elementary states x_i, which are tied to the process dynamics and parameters of the control system. We can well assert: $\forall i, : \alpha_i >> \lambda_i$

This remark means that two different time scales are present in the system: a "fast" time scale relates to the occurrence of intern events, and a "slow" time scale relates to the occurrence of faults in the system. This property will be used later for the state aggregation process.

11.3.1.3. *Equation deduced from the Markov graph*

Now, let us suppose that the system can be modeled in a Markov process like the one shown in Figure 11.3. The elementary states are lumped together according to their affiliation to a same operation mode (displaying the strongly connected graphs). We use the next notations for the generalization of the simultaneous equations:

$\mathcal{N}$: number of elementary states of the reachability graph;

n : number of macrostates;

J_i : macrostate number i with $i \in \{1..n\}$;

N_i : number of elementary states in each macrostate $J_i, i \in \{1..n\}$.

Thus, $\mathcal{N} = \sum_{k=1}^{n} N_k$;

x_{ji} : elementary state number i in the macrostate J_j with $i \in \{1..N_j\}, j \in \{1..n\}$;

α_{ij}^{k} : transition rate from the state x_{ki} to the state x_{kj} in the macrostate J_k;

λ_{ij}^{kl} : failure rate from the state x_{ki} in the macrostate J_k to the state x_{lj} in the macrostate J_l;

$q_{ji}(t)$: probability to be in the state x_{ji} at time t with $i \in \{1..Nj\}, j \in \{1..n\}$;

$Qj(t)$: probability to be in the macrostate J_j at time t. Thus, $Qj(t) = \sum_{k=1}^{N_j} q_{jk}(t)$

$p_{ji}(t)$: conditional probability to be in the state x_{ji} at time t, given that the system is in the macrostate J_j, with $i \in \{1..Nj\}, j \in \{1..n\}$. Thus, $p_{ji}(t) = \Pr[x_{ji}/J_j] = \Pr[x_{ji} \cap J_j]/\Pr[J_j] = \frac{q_{ji}(t)}{q_{j1}(t)+q_{j2}(t)+...+q_{jN_j}(t)} = \frac{q_{ji}(t)}{Q_j(t)}$

The first aim is to display a differential equation system. Its resolution gives the probabilities of being in the macrostates. Thus, the terms $Q_j(t)$ and their derivatives should appear.

Using the previous notations, we can write:

$$q_{ji}(t + dt) = \Pr[system\_in\_a\_state \neq x_{ji}\_at\_t\_and\_in\_the\_state\_x_{ji}\_at\_t + dt] + \Pr[system\_in\_the\_state\_x_{ji}\_at\_t\_and\_remains\_in\_it)]$$

that we will note:

$$q_{ji}(t + dt) = \Pr[\mathcal{A}] + \Pr[\mathcal{B}]$$

and then develop the first term:

$$\Pr[\mathcal{A}] = P[system\_in\_a\_state \neq x_{ji}\_at\_t\_in\_another\_Macro - State \neq J_j + P[system\_in\_a\_state \neq x_{ji}\_at\_t\_in\_the\_Macro - State\_J_j]$$

that we will note:

$$\Pr[\mathcal{A}] = \Pr[\mathcal{C}] + \Pr[\mathcal{D}]$$

Developing each term, we obtain:

$$\Pr[\mathcal{D}] = \sum_{k-1,k\neq i}^{N_j} q_{jk}(t).\alpha_{ki}^{j}.dt$$

$$\Pr[\mathcal{C}] = \sum_{l=1,l\neq j}^{n}\sum_{k=1}^{N_j} q_{lk}(t).\lambda_{ki}^{lj}.dt$$

$$\Pr[\mathcal{B}] = q_{ji}(t).\left[1 - \left(\sum_{k=1,k\neq i}^{N_j} \alpha_{ik}^{j} + \sum_{l=1,l\neq j}^{n}\sum_{k=1}^{N_l} \lambda_{ik}^{jl}\right).dt\right]$$

and thus:

$$\frac{q_{ji}(t+dt) - q_{ji}(t)}{dt} = \dot{q}_{ji}(t) = \sum_{k=1,k\neq i}^{N_j} q_{jk}(t).\alpha_{ki}^{j} +$$

$$\sum_{l=1,l\neq j}^{n}\sum_{k=1}^{N_j} q_{lk}(t).\lambda_{ki}^{lj} - q_{ji}(t).\left[\left(\sum_{k=1,k\neq i}^{N_j} \alpha_{ik}^{j} + \sum_{l=1,l\neq j}^{n}\sum_{k=1}^{N_l} \lambda_{ik}^{jl}\right)\right]$$

According to the definition of $Q_j(t) = \sum_{k-1}^{N_j} q_{jk}(t)$, we have:

$$\dot{Q}_j(t) = \sum_{p=1}^{N_j} \dot{q}_{jp}(t) = \sum_{p=1}^{N_j}\sum_{k=1,k\neq p}^{N_j}$$

$$q_{jk}(t).\alpha_{kp}^{j} + \sum_{p=1}^{N_j}\sum_{l=1,l\neq j}^{n}\sum_{k=1}^{N_l} q_{lk}(t).\lambda_{kp}^{lj} -$$

$$\sum_{p=1}^{N_j}\sum_{k=1,k\neq p}^{N_j} q_{jp}(t).\alpha_{pk}^{j} - \sum_{p=1}^{N_j}\sum_{l=1,l\neq j}^{n}\sum_{k=1}^{N_l} q_{jp}(t).\lambda_{pk}^{jl}$$

We can notice that:

$$\sum_{p=1}^{N_j}\sum_{k=1,k\neq p}^{N_j} q_{jk}(t).\alpha_{kp}^{j} - \sum_{p=1}^{N_j}\sum_{k=1,k\neq p}^{N_j} q_{jp}(t).\alpha_{pk}^{j} =$$

$$\sum_{p=1}^{N_j}\sum_{k=1,k\neq p}^{N_j} q_{jk}(t).\alpha_{kp}^{j} - \sum_{k=1}^{N_j}\sum_{p=1,p\neq k}^{N_j} q_{jp}(t).\alpha_{pk}^{j} =$$

$$\sum_{p=1}^{N_j}\left(\sum_{k=1,k\neq p}^{N_j} q_{jk}(t).\alpha_{kp}^{j} - \sum_{p=1,p\neq k}^{N_j} q_{jp}(t).\alpha_{pk}^{j}\right) = 0$$

It gives:

$$\dot{Q}_j(t) = \sum_{p=1}^{N_j}\sum_{l=1,l\neq j}^{n}\sum_{k=1}^{N_l} q_{lk}.\lambda_{kp}^{lj} - \sum_{p=1}^{N_j}\sum_{l=1,l\neq j}^{n}\sum_{k=1}^{N_l} q_{jp}(t).\lambda_{pk}^{jl}$$

Finally, considering the definition of the conditional probabilities $p_{ji}(t)$:

$$\dot{Q}_j(t) = \sum_{l=1,l\neq j}^{n} Q_l(t)\left[\sum_{p=1}^{N_j}\sum_{k=1}^{N_l}\lambda_{kp}^{lj}.p_{lk}(t)\right] -$$

$$Q_j(t)\left[\sum_{p=1}^{N_j}\sum_{l=1,l\neq j}^{n}\sum_{k=1}^{N_l}\lambda_{pk}^{jl}.p_{jp}(t)\right], \forall j \in [1...n]$$

Or in a condensed form:

$$\dot{Q}_j(t) = \sum_{l=1,l\neq j}^{n} \beta_l^j(t).Q_l(t),\ \forall j \in [1...n] \qquad [11.1]$$

with:

$$\begin{cases} \beta_l^j(t) = \sum_{p=1}^{N_j}\sum_{k=1}^{N_l}\lambda_{kp}^{lj}.p_{lk}(t) \quad \text{for } l \neq j \\ \beta_l^j(t) = -\sum_{p=1}^{N_j}\sum_{l=1,l\neq j}^{n}\sum_{k=1}^{N_l}\lambda_{pk}^{jl}.p_{jp}(t) \end{cases}$$

This form allows us to display the probabilities to be in the macrostates J_j, and the equivalent transitions $\beta_l^j(t)$. They are equal to a linear combination of the elementary failure rates which are weighted by temporal functions $p_{ij}(t)$.

Note that equation [11.1] is a condensed equation system deduced from the initial graph of Figure 11.3. No reduction has been used to solve it. We can nevertheless deduce the aggregated Markov graph from the initial one as shown in Figure 11.4.

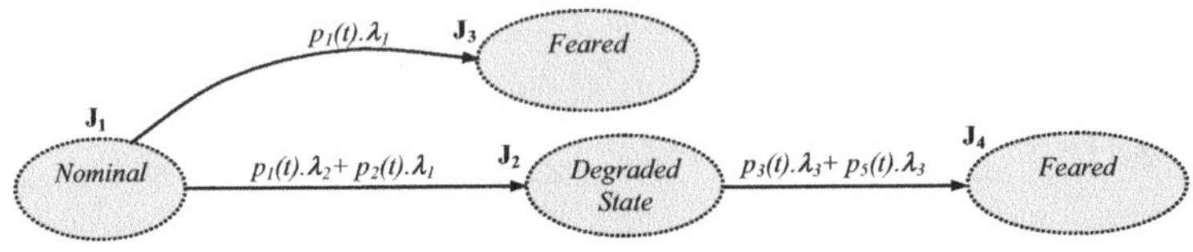

Figure 11.4. *Aggregated Markov graph*

Nevertheless, this aggregated model is not homogeneous anymore, because the transition rates $\beta_l^j(t)$ are time-dependent. This makes the system more difficult to solve. Thus, the resolution of equation [11.1] requires knowing the $\beta_l^j(t)$, and more exactly the conditional probabilities $p_{ji}(t)$, which are not easy to evaluate (without resolving the Chapman–Kolmogorov equation associated with the initial Markov process, presented in Figure 11.3).

The following section deals with a simplifying method for the differential equation [10.6] by approximations of the $p_{ji}(t)$.

11.3.1.4. *Approximations of the simultaneous differential equations*

We will use a mathematical tool which allows us to find an analytical simplification for the fundamental equation [10.6]. We will take advantage of the existence of a double time scale as stated in the previous section.

The singular perturbation method [KOK 99] allows us to identify and separate slow and fast variables of equation [11.1]. We will bring out the "standard form":

$$\left\{ \begin{array}{l} \dot{Q} = f\left(Q, p, \lambda, \tau\right) \\ \lambda.\dot{p} = g\left(Q, p, \lambda, \tau\right) \end{array} \right. \quad [11.2]$$

Thus, equation [11.1] will be split into two subsystems. The first defines the evolution of the slow variables $Q_j(t)$, and the second defines the evolution of the fast variables, $p_{ji}(t)$. $p_{ji}(t)$ were described previously, and their feature is to reach the final stationary distribution quickly. The probabilities $Q_j(t)$ of being in a functional mode are characterized by the failure rates, so we conclude that they evolve slowly. The standard form [11.2] displays the separation between slow and fast time scales (see [KOK 99]).

Now, we will describe how the standard form can be obtained from [11.1].

By differentiating the expression $p_{ji}(t).Q_j(t) = q_{ji}(t)$, we have: $\dot{q}_{ji} = \dot{p}_{ji}.Q_j + p_{ji}.\dot{Q}_j$

We substitute the terms by their expression given in [11.1]:

$$\dot{q}_{ji} = \sum_{k=1,k\neq i}^{N_l} q_{jk}.\alpha_{ki}^{j} + \sum_{l=1,l\neq j}^{n}\sum_{k=1}^{N_l} q_{lk}.\lambda_{ki}^{jl} - q_{ji}.\left(\sum_{k=1,k\neq i}^{N_j} \alpha_{ki}^{j} + \sum_{l=1,l\neq j}^{n}\sum_{k=1}^{N_l}\lambda_{ik}^{jl}\right), \forall j,i$$

and then:

$$\begin{aligned}\dot{p}_{ji}.Q_j = & -p_{ji}.\sum_{p=1}^{N_j}\sum_{l=1,l\neq j}^{n} Q_l\sum_{k=1}^{N_l}.p_{lk}.\lambda_{kp}^{lj} \\ & +Q_j.p_{ji}.\sum_{p=1}^{N_j}\sum_{l=1,l\neq j}^{n}\sum_{k=1}^{N_l}.p_{jp}.\lambda_{pk}^{jl} + \\ & Q_j.\sum_{k=1,k\neq i}^{N_j} p_{jk}.\alpha_{ki}^{j} + \sum_{l=1,l\neq j}^{n} Q_l\sum_{k=1}^{N_l} p_{lk}.\lambda_{ki}^{jl} - \\ & Q_j.p_{ji}\left(\sum_{k=1,k\neq i}^{N_j}\alpha_{ik}^{j} + \sum_{l=1,l\neq j}^{n}\sum_{k=1}^{N_l}\lambda_{ik}^{jl}\right)\end{aligned}$$

and finally, we divide this expression by $Q_j(t)$:

$$\dot{p}_{ji} = -p_{ji}.\sum_{p=1}^{N_j}\sum_{l=1,l\neq j}^{n}\frac{Q_l}{Q_j}\sum_{k=1}^{N_l}.p_{lk}.\lambda_{kp}^{lj} + p_{ji}.\sum_{p=1}^{N_j}\sum_{l=1,l\neq j}^{n}\sum_{k=1}^{N_l}.p_{jp}.\lambda_{pk}^{jl} +$$

$$\sum_{k=1,k\neq i}^{N_j} p_{jk}.\alpha_{ki}^{j} + \sum_{l=1,l\neq j}^{n}\frac{Q_l}{Q_j}\sum_{k=1}^{N_l}p_{lk}.\lambda_{ki}^{jl} - p_{ji}\left(\sum_{k=1,k\neq i}^{N_j}\alpha_{ik}^{j} + \sum_{l=1,l\neq j}^{n}\sum_{k=1}^{N_l}\lambda_{ik}^{jl}\right)$$

[11.3]

We can see that equations [11.1] and [11.3] represent the expected standard form. We have just to set:

$\lambda = Max(\lambda_{ij}^{kl}), (j,l) \in \{1..N_j\}, (i,k) \in \{1..n\}$

$\lambda_{ji}^{kl} = \lambda.\delta_{ji}^{kl}, (j,l) \in \{1..Nj\}, (i,k) \in \{1..n\}, \delta_{ji}^{kl} \in]0,1]$

And we choose a new time scale: $d\tau/dt = 1/\lambda$

Then:

$$\begin{cases} \frac{dQ_j}{d\tau} = \sum_{l=1,l\neq j}^{n} Q_l\left[\sum_{p=1}^{N_j}\sum_{k=1}^{N_l}\delta_{kp}^{li}.p_{lk}\right] - Q_j\left[\sum_{p=1}^{N_j}\sum_{l=1,l\neq j}^{n}\sum_{k=1}^{N_l}\delta_{pk}^{jl}.p_{jp}\right] \\ \lambda.\frac{dp_{ji}}{d\tau} = \lambda\left[\begin{array}{c} -p_{ji}\sum_{p=1}^{N_j}\sum_{l=1,l\neq j}^{n}\frac{Q_l}{Q_j}.\sum_{k=1}^{N_l}p_{lk}.\delta_{kp}^{lj} + p_{ji}.\sum_{p=1}^{N_j}\sum_{l=1,l\neq j}^{n}\sum_{k=1}^{N_l}p_{jp}.\delta_{pk}^{jl} + \\ \sum_{l=1,l\neq j}^{n}\frac{Q_l}{Q_j}\sum_{k=1}^{N_l}p_{lk}.\delta_{ki}^{jl} - p_{ji}.\sum_{l=1,l\neq j}^{n}\sum_{k=1}^{N_l}\delta_{ik}^{jl} \end{array}\right] + \\ \sum_{l=1,l\neq j}^{N_j} p_{jk}.\alpha_{ki}^{j} - p_{ji}.\sum_{l=1,l\neq j}^{N_j}\alpha_{ik}^{j} \end{cases}$$

[11.4]

The first equation of equation system [11.4] gives the evolution of the slow behavior relative to the probabilities of being in the macrostates. These probabilities depend on the failure rates. The second subsystem gives the evolution of the fast behavior, because the set of the probabilities $p_{ij}(t)$ defines the internal system dynamics through a set of statistical parameters. Moreover, they are independent of the failure occurrences.

The idea is to characterize the dynamics with a set of variables which will be injected in a reduced Markov model. It is typically a problem of dynamic reliability.

The simplification of equation system [11.4] is based on the reduction of its order. The principle of the singular perturbations is to suppose that the fast variables ($p_{ij}(t)$) reach their asymptotic limit quickly (subject to their existence), in comparison with the slow variables $Q_j(t)$. The reduced model is composed of the first equation of system [11.4] in which the $p_{ji}(t)$ are replaced by their limit.

The equation system [11.4] also reveals the existence of a small scalar λ to be neglected. The simplifying method consists of setting $\lambda = 0$ (zero-order approximation in singular perturbation method [KOK 99]). The dimension of the state space changes from $\mathcal{N}$ to n. The second differential equation of system [11.4] becomes an algebraic equation easy to resolve (the variables coated with a bar are solutions of the equation resulting from the approximation), $0 = g\left(\overline{Q}, \overline{p}, 0, \tau\right)$, in other words for each subgraph J_j:

$$0 = \sum_{k=1,k\neq i}^{N_j} \overline{p}_{jk}.\alpha_{ki}^{j} - \overline{p}_{ji}. \sum_{k=1,k\neq i}^{N_j} \alpha_{ik}^{j} \;, \forall i \in [1, Nj] \qquad [11.5]$$

When solving this system, the limits are not unique, that is why we add the following constraints on the probabilities:

$$\forall j \in [1, n] \qquad , \sum_{i=1}^{N_j} \overline{p}_{ji} = 1 \qquad [11.6]$$

The reduced Markov process consists of n states relating to n functional or dysfunctional modes, which were previously identified.

The probility of being in a macrostate at time t is deduced after solving the "slow" equation: $\dot{\overline{Q}} = f\left(\overline{Q}, \overline{p}, \lambda, \tau\right)$, where $\overline{p}$ is the set of the stationary probabilities.

NOTES.–

– The resolution of equation [11.5] under constraint [11.6] gives a unique solution for each conditional probability. It is provided by the assumption on the structure of the subgraphs which are strongly connected: each J_j constitutes an ergodic class, and then the resolution of [11.5]–[11.6] corresponds to the search of the limit distribution of the conditional probabilities, and it is independent of the initial values of $p_{ji}(t)$.

– The possibility of system repair induces the presence of an intermediate time scale between the fast dynamic (intrinsic to the evolution of the system), and the slow dynamic suited to the failure occurrence. Nevertheless, the described method allows for reducing the size of the state space, by removing the faster dynamic, so that the system is featured by two dynamics instead of three. In principle, the Markov graph structure remains identical after adding reparation arcs, providing thereby a good overview of the whole system and its functioning modes.

– In this section, the hypothesis providing a perfect splitting of the dynamics and the displaying of the macrostates is based solely on the transition between two functioning modes by failure occurrence. In practice, the control system can switch, for example, between two functioning modes on user request or if the system needs to change reconfiguration mode. This possibility was not taken into consideration, because it involves the appearance of more or less "fast" transient states between macrostates. These states must be analyzed one by one, in order to include them if possible in one of the sets, or to assign a transition rate of the same order as the failure rates, provided that the request frequency allows such an approximation. Nevertheless, this instance could be handled with the proposed method in [BOB 86].

11.3.2. *Toward a direct establishment of a reduced Markov graph*

The application of this method requires knowing the final stationary distribution of the conditional probabilities. Nevertheless, in practice, the initial PN does not need to be randomized by assigning exponential probability law to the transitions (a transition rate α_i for each transition T_i of the PN). The way to determine the limit of the $p_{ji}(t)$ consists of processing each subgraph independently (strongly connected component), i.e. the system is in a given functional mode and then the output arcs corresponding to failure occurrence must be deleted. Once the system is in a given configuration, a Monte-Carlo simulation allows us to deduce the $\bar{p}_{ji}$. In this case, these probabilities are similar to the average marking in each state of the reachability graph.

Moreover, the simulation time remains acceptable: on the one hand, the subgraphs are simulated one by one, and on the other hand, only the behavioral model is simulated in the absence of any failure. So, there is no longer a huge computer simulation time due to multiple time scales.

11.4. Example

As an illustration of this method, we propose the didactical example [SCH 03] of an elementary control system. A control interpreted PN composed of places P_1, P_2, P_3 and P_4 modelizes the functional architecture of a system in nominal operating mode. To these places are associated actions and operations not described here. Events specific to the controlled part are associated with transitions T_1, T_2, T_5, T_4 and T_7. Their firings allow the control function to go from a control operation to another. This functional architecture is dependent on the hardware used for its implementation. The function evolving at its own rhythm, exchanges information with its environment by means of sensors and actuators. The function is assumed to be implemented on a main computer C_1. The specification of the control system in PN being made, the designer needs access to the assessment of the reliability of the system to, if necessary, propose improvements if the requirements are not satisfied. These improvements may be made on functional level as well as on the hardware.

The quantitative analysis might be made on a global PN model. It would comprise a functional part specifying the behavior of the control system and a dysfunctional part describing the failure occurrences of the components such as sensors, actuators and computers. The interaction of both parts would describe the behavior of the control system in the presence of failures.

11.4.1. *Failure modeling*

The stochastic PN model is used, the firing of a transition being associated with the failure of a component. We suppose that the stochastic variables associated with operation times are exponentially distributed.

The interactions between the functional and dysfunctional models are represented by the mean of immediate transitions. An example is given in Figure 11.5.

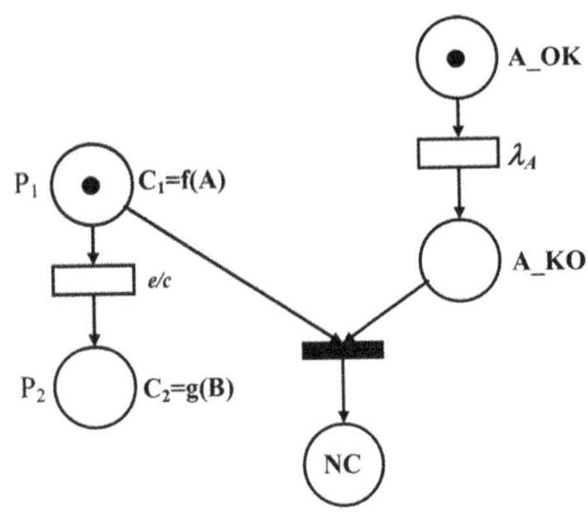

Figure 11.5. *Failure modeling and interaction with the control*

From information measured by the sensor A, the operation assigned to P_1 calculates a set point to be sent to the actuator C_1. The states of the sensor are represented by places A_OK and A_KO for operating and failure states, respectively.

The firing of a stochastic transition on a failure occurrence puts the system in the non-conformity NC state if the function is using this information at the same time. Formally, the simultaneous marking of places P_1 and A_KO enabled an immediate transition leading the system to switch to the dysfunctional state.

Let us remark that the PN in Figure 11.5 is neither a stochastic PN nor an interpreted PN if we refer to the definitions of these two models. It can be considered to be an extended stochastic PN to which an interpretation would be added to immediate transitions. Formally, this model has not been defined but it may be compared to the hybrid stochastic automaton presented in the first book of this series [AUB 15] in which the underlying graph is replaced by the reachability graph of the underlying autonomous PN in Figure 11.5. The simulation of the model can then be imagined but runs the risk of being highly time-consuming due to the cohabitation of very different dynamics in the temporal evolution of the different variables. That is why we will propose the alternative approach of an analytical evaluation by the use of the singular perturbation theory.

In the example, the reliability study will be based on the study of the probability of being in the place NC at any time.

11.4.2. *Study of the different functional and hardware solutions*

To implement the control system of our example, four solutions have been studied in order to consider several kinds of fault tolerance mechanisms. In the first solution, no functional redundancy is used and no hardware redundancy is planned. In the second solution, a passive hardware redundancy of the computer is introduced. In the third solution, a software reconfiguration is proposed. It allows the system to tolerate a failure in sensor A. In case of this failure, the systems run a degraded operating mode. In the fourth solution, both strategies are implemented: the redundancy of the computer and the functional reconfiguration on a sensor failure.

Figure 11.6 shows the representation of solution four; three parts are to be distinguished:

– the dysfunctional part: it is formalized by stochastic PNs to represent the whole set of component failures and allows performances assessment in terms of dependability. The failures of actuator A and computers C_1 and C_2 are considered;

– two functional parts describing the control algorithm of the system implemented on both computers. The operations of the affected places are not described. One of these two parts is active at any one time according to the state of the computers.

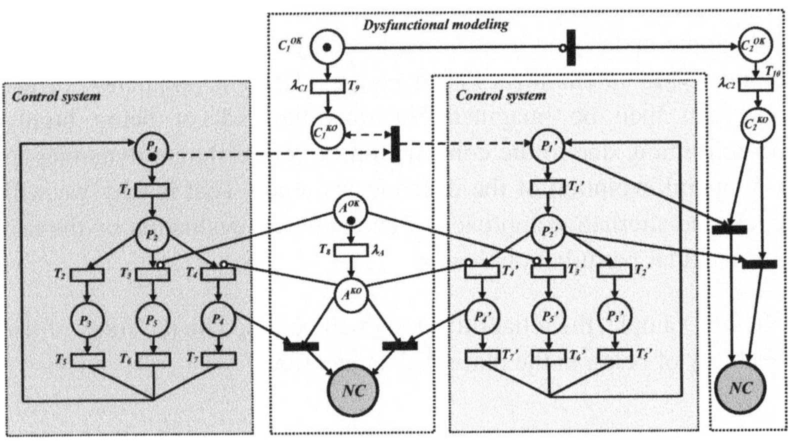

Figure 11.6. *PN model of the control system*

The nominal operation is represented by the sequences $\{T_2, P_3, T_5\}$ (respectively, $\{T_2', P_3', T_5'\}$) and $\{T_4, P_4, T_7\}$ (respectively, $\{T_4', P_4', T_7'\}$), the sequences $\{T_3, P_5, T_6\}$ (respectively, $\{T_3', P_5', T_6'\}$) modelize the degraded operating mode. The failure of the sensor A substitutes the sequence $\{T_4, P_4, T_7\}$ (respectively, $\{T_4', P_4', T_7'\}$) by $\{T_3, P_5, T_6\}$ (respectively, $\{T_3', P_5', T_6'\}$). However, if the sensor fails when the system uses its information (token in place P_4), then the system enters the non-conforming state (token in NC).

A classical analytical study for the assessment of dependability implies the development of the marking (reachability) graph in order to deduce a Markov process. In our example, the transitions of the control interpreted PN part are not stochastic, their firing being linked to deterministic events. Their purpose is to synchronize the control system with the controlled part. To simplify the demonstration, we will transform these transitions into stochastic ones by forgetting their interpretation (the event $\mu(T_i)$ and predicate $\psi(T_i)$) and assigning them a constant transition rate α_i (exponential distribution) as shown in Figure 11.7.

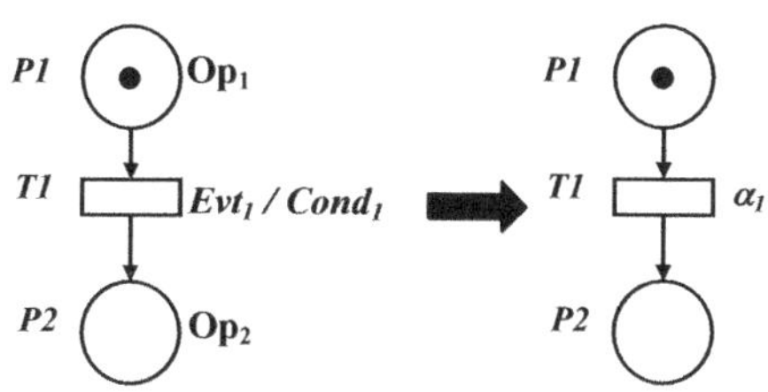

Figure 11.7. *"Stochastization" of the control transitions*

As the behavior of the control system is governed by a much faster dynamic than the dysfunctional behavior, we put forward the hypothesis that:

$$\alpha_i >> \lambda_j \qquad \forall i, j$$

The knowledge of the marking graph for each of the four aforementioned solutions leads us to know the corresponding Markov graph. Concerning the fourth model, it is composed of 17 states (Figure 11.8).

In this graph, two dynamics coexist due to the presence of fast transition rates α_i and slow transition rates λ_j. Let us now uncouple these dynamics to reduce the size of the Chapman–Kolmmogorov equation by the identification of the state sets constituting strongly connected components.

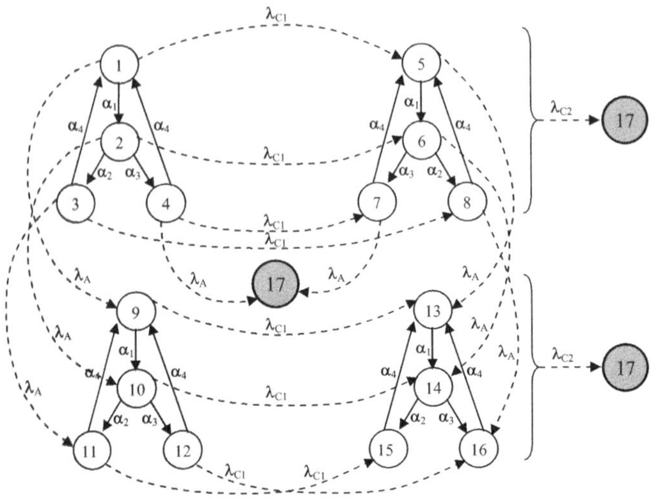

Figure 11.8. *Markov graph of the fourth model*

The first step consists of the elimination of the arcs related to slow transition rates. Thus, four strongly connected components appear in addition to the trivial one reduced to the absorbing state 17 (non-conformity):

– component 1 : states {1, 2, 3, 4};

– component 2 : states {5, 6, 7, 8};

– component 3 : states {9, 10, 11, 12};

– component 4 : states {13, 14, 15, 16}.

It appears that each component identifies itself to one of the different functioning modes:

– component 1 ↔ nominal mode MN (no failure);

– component 2 ↔ degraded mode MD1 (hardware reconfiguration on computer C_2 after a failure of C_1);

– component 3 ↔ degraded mode MD2 (software reconfiguration after a failure of the sensor A);

– component 4 ↔ degraded mode MD3 (hardware and software reconfigurations).

Each subgraph only contains fast transition rates relative to functioning modes of the control algorithm. They are linked to each other by failure transitions leading the system to switch from the nominal mode to a degraded one or to NC mode.

From these considerations, we can establish the reduced Markov graph, the macrostates (in which we search for the probabilities) corresponding to the identified functioning modes. The equivalent transition rates are coming from the C-K reduced state equation.

They are equal to linear combinations of the failure rates between macrostates weighted by the conditional probabilities (asymptotic) of being in the elementary state upstream the considered failure rates.

The aggregated Markov graph is presented in Figure 11.9.

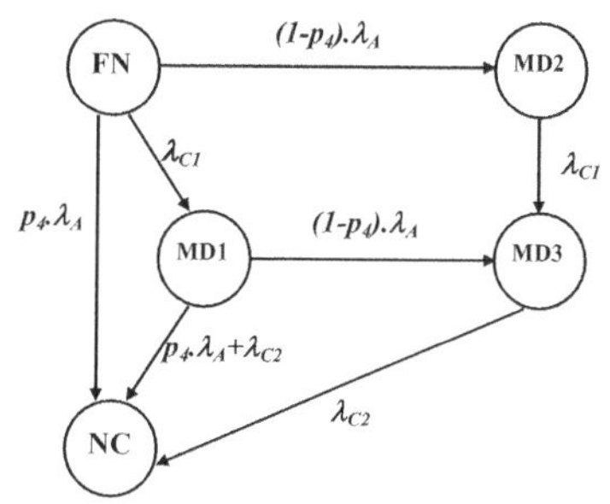

Figure 11.9. *Aggregated Markov graph*

The coefficient p_4 weighting the failure rate λ_A is the asymptotic conditional probability for the system being in state 4 knowing that it is in nominal functioning mode. Effectively, a failure of sensor A may have different consequences according to the failure time occurrence:

– if the failure occurs when the system operates on the sensor information, that is to say if place P_4 is marked, then the systems enter the NC state;

– otherwise, a software reconfiguration mechanism puts the system into degraded mode MD2.

Therefore, the probability of the system going into non-conformity due to a sensor failure is directly linked to the probability of place p_4 being marked.

The weighting term p_4 is also present on the failure rate λ_A between the macrostates MD1 and NC. Inversely, the probability of the system entering the software reconfiguration modes MD2 or MD3 depends on the weighting term $(1 - p_4)$.

The calculus of the whole set of asymptotic conditional probabilities for each strongly connected component requires the extraction from the global transition matrix of the four submatrices specific to each component after the elimination of the slow transition rates (order 0 approximation of the singular perturbation theory). However, such a calculus does not hold because the transition rates arbitrarily introduced have no physical significance.

However, note that the weighting coefficients in the transition rates of the aggregated graph are dependent on the marking of the PNs modeling the macrostates. These considerations point to a solution for a direct evaluation of the weighting coefficients from the PN without using artificial "stochastization".

11.4.3. *Evaluation of the weighting coefficients from the Petri nets*

The direct connection between the weighting coefficients appearing in the expression of conditional probabilities in steady state and the probability of the PN marking is only possible if the PN is safe. On the contrary, the transition rates of the Markov graph are dependent on the current marking and on the firing rates of the considered PN.

We suppose in the following that the PN is a safe, legitimate hypothesis in case of control interpreted PNs used here.

Let us recall that the marking process is ergodic if it converges toward the same finite limit in temporal and probabilistic mean value. This limit is denoted by $\overline{M}$ vector of mean marking in steady state.

The marking processes relative to each functioning mode will be supposed as in the following. The calculus of weighting coefficients can then be made analytically in case of generalized stochastic PNs; in other cases, these coefficients may be evaluated by simulation.

Let us return to our example (section 11.4.2). The fourth solution comprise five macrostates, four of them corresponding to a nominal or degraded mode. These four macrostates play a role in the calculus of the weighting coefficients because they correspond to transitory state classes. To evaluate these weighting coefficients, we propose the following procedure:

– for each functioning mode, nominal or degraded, build the PN that models the behavior of the system in this mode;

– evaluate the mean marking in each place of these partial PNs, supposing that the systems stay in the mode indefinitely, in other words supposing that no failure occurs. In the case of control interpreted PNs in which the global behavior is simultaneously characterized by the evolution of the discrete part as well as the continuous one, the mean markings may be evaluated by a hybrid simulation;

– for each partial PN, identify the places whose marking probability intervenes in the calculus of equivalent rates.

The fourth solution is hence composed of four partial PNs. Figure 11.10 shows the PNs representing the nominal mode (left) and the degraded mode 3 with hardware and software reconfigurations (right). The PN corresponding to the degraded mode 1 (hardware reconfiguration) is identical to the nominal mode (same names of places and transitions with prime superscript). In the same way, the PN of degraded mode 2 (software reconfiguration) is identical to this of mode 3.

By generating the marking graphs of these partial PNs, the strongly connected components of the initial PN are recognized.

This connectivity property must be systematically verified. It translates into a structural viewpoint of the liveness property of the associated partial PN. We suppose, in addition, that the liveness is maintained after the addition of the interpretation (interpreted PN); in other words, the modeled system has all the "good" properties. This property is acquired when the quantitative study follows a verification of the model which allows us to guarantee that no deadlock is present.

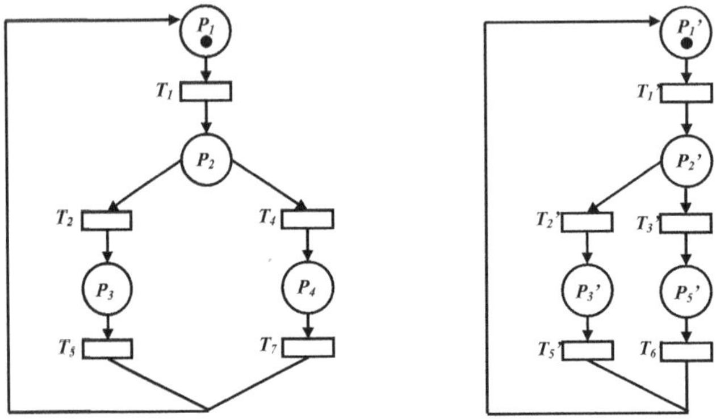

Figure 11.10. *Two examples of sub-PNs*

When satisfying the ergodic property, Markov partial graphs have a unique limit distribution of probabilities independent of the initial conditions.

The analogy between the asymptotic distribution of probabilities and the marking process allows us to write for each functioning mode j,

$$\sum_{i/P_i \in Mode\_j} \overline{M}(P_i) = 1$$

conditionally of the uniqueness of the token describing the evolution of the control system in each mode.

The use of simulation to evaluate the mean marking in each functioning mode does not pose any time problem as they do not contain multiple dynamics.

11.4.4. *Conclusion*

In this example, we have shown that it is possible to reduce the size of a system in the purpose of its dependability assessment. The singular perturbation method allows for the uncoupling of variables highly different in dynamics. A Markov macromodel of reduced size representing a slow dynamic is obtained to find an analytical solution. A preliminary study of the modes corresponding to each macrostate is performed under the form of PNs representing the fast dynamics. The simulation of these PNs gives access to the mean markings whose knowledge intervenes in the calculus of the macromodel transition rates. Subject to the existence of a limit distribution of the marking process, these simulations do not pose any difficulty and can be made with any existing PN software tool.

12

Extensions of SPN

12.1. Introduction

The stochastic Petri nets (PNs) we have presented in previous chapters model the firing random time of a transition with an exponential distribution. Other types of stochastic PNs have been developed to take other distribution types into account and extend the modeling and analysis power of systems including random phenomena in their behavior.

The best known of them are (by increasing generalization order):

– the generalized stochastic Petri nets (GSPNs) [CHI 93, AJM 84] considering two transition types: stochastic transitions with firing times exponentially distributed and immediate transitions instantaneously fired after being enabled;

– the deterministic and stochastic Petri nets (DSPNs) [LIN 98, AJM 86] considering three transition types: stochastic transitions with firing times exponentially distributed, immediate transitions and deterministically timed transitions;

– the extended stochastic Petri nets (ESPNs) [DUG 84] considering general (non-exponential) distributions for the transition firing time, except for concurrent transitions (see section 1.3.1) that are restricted to exponential distributions;

– the Markov regenerative stochastic Petri nets (MRSPNs) [CHO 94] considering general distributions but among transitions enabled in each marking at most one of them can have a generally distributed firing time, the other ones being immediate or exponentially distributed. This last type of SPN is the most general one, and it includes the three previous ones.

For any of these SPN extensions, the set of transitions T is partitioned into two subsets T_{im} (immediate) and T_{te} (timed) such that $T = T_{im} \cup T_{te}$, $T_{im} \cap T_{te} = \varnothing$, the subset T_{te} of timed transitions potentially containing three types:

– with deterministic firing time;

– with exponentially distributed firing time;

– with firing time distributed according to any other probabilistic law than exponential (subset transition denoted by T_g).

12.2. Relationship between stochastic Petri nets and stochastic processes

The analysis of random phenomena of a system involves the joint use of stochastic PNs and stochastic processes.

The relation between a stochastic PN and its associated stochastic process is established by means of a marking graph of the PN:

– the reachability graph of an SPN is isomorph to a continuous time Markov chain;

– the reachability graph of a GSPN is also isomorph to a continuous time Markov chain;

– the reachability graph of an ESPN is isomorph to a semi-Markov process (under some constraints);

– the reachability graph of a DSPN is isomorph to a semi-regenerative process (under some constraints);

– the reachability graph of an MRSPN is isomorph to a semi-regenerative process (under some constraints).

12.3. The transition firing policy

Before studying the way in which to define the stochastic processes associated with the SPNs, we will present some notions about the transition firing policy. This stochastic process is essentially dependent on the choice of such a policy.

The transition firing policy comprises two aspects: the selection of the transition to be fired for a given marking enabling several transitions and the conditioning on the past history to keep track of the past behavior when a transition is fired.

Two different methods are distinguished for the selection of the transition to be fired [AJM 89]:

– the race: it induces the firing of the transition whose the associated variable is statistically the smallest;

– the preselection: firing probabilities are calculated from weights *a priori* attributed to the transitions either directly in the PN or at the reachability graph level.

To condition PN behavior on the past history when a transition is fired, the remaining time to fire each transition can be defined as the time from the current moment to the moment when it should be really fired. Based on this remaining firing time, three methods are possible [AJM 89]:

– resampling: after the firing of a transition among a set of transitions enabled by a marking, all the distribution functions of the remaining firing time for non-fired transitions are reset to zero;

– age memory: the remaining time to fire a transition is kept even if the transitions are disabled;

– enabling memory: the remaining time to fire is only kept as long as the transition remains enabled.

12.4. Associated stochastic processes

The reachability graph of an SPN contains two types of states:

– unstable states in which at least one immediate transition is fireable;

– tangible states in which only timed transitions are fireable.

In unstable states, immediate transitions are fired without time duration. These states have no interest in the study of system behavior; they will be eliminated and the reachability graph is reduced to the tangible states.

The removal of unstable states is made according to the commutation distribution associated with the immediate transitions in conflict. Hence, it is possible to define the probability for the firing sequence from a tangible state M_i to another one M_j ($M_i \overset{S}{\rightarrow} M_j$ all transitions of S being immediate):

$$\Pr(M_i, M_j) = \prod_{T_l \in S} p_l \qquad [12.1]$$

where p_l is the occurrence probability associated with the immediate transitions T_l fired to evolve in the sequence S.

Thereby, we obtain a reachability graph containing only tangible states between which only timed transitions take place.

Let us now see the influence of the temporal memories on the associated stochastic processes.

12.4.1. *Temporal memory based on resampling*

In the case where this method is used, the stochastic process does not depend on the prior evolution of the PN. The stochastic process associated with the PN is a semi-Markov process, each marking being a regeneration point.

The probability for a system to evolve from the marking M_i toward the marking M_j during the time interval $[0, t]$ is:

$$Q_{ij}(t) = \int_0^t \prod_{k \neq j} (1 - F_{ik}(u)) dF_{ij}(u) \qquad [12.2]$$

where $F_{ij}(t)$ is the cumulative distribution function of the firing time of the transition from the marking M_i to the marking M_j.

Equation [12.2] allows the determination of the kernel of the semi-Markov process associated with the PN knowing the distribution function of the firing times of the transitions leading the PN to evolve from the actual state M_i to the future state M_j.

If the firing times of the transitions are exponentially distributed, the semi-Markov process is reduced to a continuous time Markov chain.

12.4.2. *Temporal memory based on age memory or on enabling memory*

When considering techniques of temporal memory based on age memory or on enabling memory, the stochastic process depends on the previous evolution of the PN. It is not possible to conclude on the stochastic process type without knowing the type of distribution function of the transition firing times.

A timed transition $T_j \in T_t$ whose the firing time is exponentially distributed is known as an exponential transition.

A timed transition $T_j \in T_t$ whose the firing time is distributed according to general distribution function (not exponential) is known as a general transition.

In addition to this distinction between exponential and general transitions, three classes of transitions must be distinguished:

DEFINITION 12.1.– A timed transition T is known as an exclusive transition in a marking M if T is the only transition fireable (enabled and allowed) from the marking M.

DEFINITION 12.2.– A timed transition T, fireable from a marking M, is known as a competitive transition if the marking M allows the firing of another transition T' and the firing of T' prohibits the firing of T.

DEFINITION 12.3.– A timed transition T, fireable from a marking M, is known as a concurrent transition if the marking M allows the firing of another transition T' and the firing of T' does not prohibit after the firing of T.

The stochastic process associated with an SPN is a semi-regenerative process if:

– there is at most one general transition fireable from each marking;

– the distribution function of the general transition is independent of the marking.

REMARK 12.1.– The semi-regenerative process is reduced to a semi-Markov process if there is no exponential transition in concurrence with the general transitions for all the accessible markings in the PN.

The semi-regenerative process is reduced to a continuous time Markov chain if the PN contains only exponential transitions.

12.4.3. *Stochastic process underlying a stochastic PN*

The generalized approach to obtain the stochastic process underlying a stochastic PN was developed in the works of Trivedi and Choi [CHO 94, CHO 93a, CHO 93b]. This approach is presented hereafter.

Let M_i be a state of the set Ω of tangible states. We will denote by $\mathcal{G}(M_i)$ the set of general transitions and by $\mathcal{E}(M_i)$ the set of exponential transitions fireable from M_i.

For each state M_i, one of the following cases is possible:

– there is no general transition fireable from M_i, $\mathcal{G}(M_i) = \varnothing$, so it can be defined: $\Lambda_i = \sum_{M_j \in \Omega} \lambda_{ij}$

λ_{ij} being the firing rate of the exponential transition, it allows us to go from M_i to M_j;

– there is one general transition fireable from M_i, $\mathcal{G}(M_i) = \{T_g\}$. In this case, $\Omega(M_i)$ is defined as the state set in which the stochastic process may be before any state change of its embedded Markov chain occurs.

The state change of the embedded Markov chain occurs either if the general transition T_g is fired or if a competitive exponential transition is fired. The stochastic process in $\Omega(M_i)$ is a subordinate continuous time Markov chain having a generator matrix $\mathbb{A}(M_i)$ defined as follows:

– if $M_j \in \Omega(M_i)$, the firing rate from M_j to $M_{j'} \in \Omega$ is given by: $\lambda_{jj'}$;

– if $M_j \notin \Omega(M_i)$, then: $\lambda_{jj'} = 0, \forall M_{j'} \in \Omega$.

Let $\Omega_\varepsilon(M_i)$ be the set of accessible states from M_i (not necessarily in direct mode) by the firing of a competitive exponential transition, and $\Omega_g(M_i)$ the set of states accessible from M_i (not necessarily in direct mode) by the firing of the general transition.

The kernel of the stochastic process associated with a stochastic PN with general transitions is given by the following theorem [CHO 94, CHO 93a]:

THEOREM 12.1.– The kernel of the stochastic process is given by:

1) For a state M_i such that $\mathcal{G}(M_i) = \varnothing$ (no general transition enabled), the state change of the embedded Markov chain is due to the firing of an exponential transition:

$$Q_{ij}(t) = \begin{cases} 0 & \Lambda_i = 0 \\ \frac{\lambda_{ij}}{\Lambda_i}(1 - e^{\Lambda_i t}) & \Lambda_i > 0 \end{cases}$$

2) For a state M_i such that $\mathcal{G}(M_i) = \{T_g\}$, the state change of the embedded Markov chain is due to either the firing of the general transition or to its disabling by the firing of a competitive exponential transition:

– if $M_j \in \Omega_\epsilon(M_i)$ and $M_j \notin \Omega_g(M_i)$ (accessible marking by a competitive exponential transition):

$$Q_{ij}(t) = \left[e^{\mathbb{A}(M_i).t}\right]_{ij}(1 - F_d(t)) + \int_0^t \left[e^{\mathbb{A}(M_i).\tau}\right]_{ij} dF_d(\tau)$$

– if $M_j \notin \Omega_\epsilon(M_i)$ and $M_j \in \Omega_g(M_i)$ (accessible marking by the general transition):

$$Q_{ij}(t) = \sum_{M' \in \Omega(M_i)} \int_0^t \left[e^{\mathbb{A}(M_i).\tau}\right]_{MM'} .dF_d(\tau).P(M', M_j)$$

– if $M_j \in \Omega_\epsilon(M_i)$ and $M_j \in \Omega_g(M_i)$ (accessible marking by either the general transition, or by a competitive exponential transition):

$$Q_{ij}(t) = \left[e^{\mathbb{A}(M_i).t}\right]_{ij} (1 - F_d(t)) + \int_0^t \left[e^{\mathbb{A}(M_i).\tau}\right]_{ij} dF_d(\tau) +$$

$$\sum_{M' \in \Omega(M_i)} \int_0^t \left[e^{\mathbb{A}(M_i).\tau}\right]_{MM'} .dF_d(\tau).P(M', M_j)$$

– if $M_j \notin \Omega_\epsilon(M_i)$ and $M_j \notin \Omega_g(M_i)$ (non-accessible marking from M_i at the next state change of the embedded Markov chain):

$$Q_{ij}(t) = 0, t \geqslant 0.$$

If the general transition T_g fireable from the marking M_i has a fixed firing time, it is known as a deterministic transition. Theorem 4.1 is replaced by the following theorem [CHO 93a, CHO 93b]:

THEOREM 12.2.– The kernel $Q(t) = [Q_{ij}(t)]$ of the stochastic process of a stochastic and deterministic PN is given by:

1) for a state M_i, such that $\mathcal{G}(M_i) = \varnothing$

$$Q_{ij}(t) = \begin{cases} 0 & \Lambda_i = 0 \\ \frac{\lambda_{ij}}{\Lambda_i}(1 - e^{\Lambda_i t}) & \Lambda_i > 0 \end{cases}$$

2) for a state M_i, such that $\mathcal{G}(M_i) = \{T_g\}$, with δ the firing time of the transition T_g:

– if $M_j \in \Omega_\epsilon(M_i)$ and $M_j \notin \Omega_g(M_i)$:

$$Q_{ij}(t) = \begin{cases} \left[e^{\mathbb{A}(M_i).t}\right]_{ij} & t < \delta \\ \left[e^{\mathbb{A}(M_i).\delta}\right]_{ij} & t \geqslant \delta \end{cases}$$

– if $M_j \notin \Omega_\epsilon(M_i)$ and $M_j \in \Omega_g(M_i)$:

$$Q_{ij}(t) = \begin{cases} 0 & t < \delta \\ \sum_{M' \in \Omega(M_i)} \left[e^{\mathbb{A}(M_i).\delta}\right]_{ij} .P(M', M_j) & t \geqslant \delta \end{cases}$$

– if $M_j \in \Omega_\epsilon(M_i)$ and $M_j \in \Omega_g(M_i)$:

$$Q_{ij}(t) = \begin{cases} \left[e^{\mathbb{A}(M_i).\delta}\right]_{ij} & t < \delta \\ \left[e^{\mathbb{A}(M_i).\delta}\right]_{ij} + \sum_{M' \in \Omega(M_i)} \left[e^{\mathbb{A}(M_i).\delta}\right]_{ij} .P(M', M_j) & t \geqslant \delta \end{cases}$$

– if $M_j \notin \Omega_\epsilon(M_i)$ and $M_j \notin \Omega_g(M_i)$

$$Q_{ij}(t) = 0, t \geqslant 0.$$

Theorems 12.1 and 12.2 allow for the determination of the kernel of the stochastic process underlying a stochastic PN knowing the distribution functions of the firing time of transitions.

The random behavior of an SPN is identical to that of the underlying stochastic process. In steady states, the study of a stochastic process requires the determination of the embedded discrete time Markov chain. The matrix of the transition probabilities of the embedded discrete time Markov chain is obtained [CHO 94, CHO 93a, CHO 93b] from the kernel given by theorems 12.1 and 12.2 and will be presented in the next section.

12.4.4. *Embedded Markov chain of the stochastic process*

COROLLARY 12.1.– The transition probability matrix $P = [p_{ij}]$ of the embedded discrete time Markov chain is given by:

1) For a state M_i, such that $\mathcal{G}(M_i) = \varnothing$

$$P_{ij} = \begin{cases} 0 & \Lambda_i = 0 \\ \frac{\lambda_{ij}}{\Lambda_i} & \Lambda_i > 0 \end{cases}$$

2) For a state M_i, such that $\mathcal{G}(M_i) = \{T_g\}$:

– if $M_j \in \Omega_\epsilon(M_i)$ and $M_j \notin \Omega_g(M_i)$:

$$P_{ij} = E\left[\left[e^{\mathbb{A}(M_i).t}\right]_{ij}\right]$$

– if $M_j \notin \Omega_\epsilon(M_i)$ and $M_j \in \Omega_g(M_i)$:

$$P_{ij} = \sum_{M' \in \Omega(M_i)} E\left[\left[e^{\mathbb{A}(M_i).\delta}\right]_{ij}\right]_{MM'} .P(M', M_j)$$

– if $M_j \in \Omega_\epsilon(M_i)$ and $M_j \in \Omega_g(M_i)$:

$$P_{ij} = E\left[\left[e^{\mathbb{A}(M_i).t}\right]_{ij}\right] + \sum_{M' \in \Omega(M_i)} E\left[\left[e^{\mathbb{A}(M_i).\tau}\right]_{MM'} .P(M', M_j)\right]$$

– if $M_j \notin \Omega_\epsilon(M_i)$ and $M_j \notin \Omega_g(M_i)$

$$P_{ij} = 0$$

COROLLARY 12.2.– The transition probability matrix of the embedded Markov chain of a stochastic process of stochastic and deterministic PNs is given by:

1) For a state M_i, such that $\mathcal{G}(M_i) = \varnothing$:

$$P_{ij} = \begin{cases} 0 & \Lambda_i = 0 \\ \frac{\lambda_{ij}}{\Lambda_i} & \Lambda_i > 0 \end{cases}$$

2) For a state M_i, such that $\mathcal{G}(M_i) = \{T_g\}$ with the firing time δ of the transition T_g:

– if $M_j \in \Omega_\epsilon(M_i)$ and $M_j \notin \Omega_g(M_i)$:

$$P_{ij} = \left[e^{A(M_i).\delta}\right]_{ij}$$

– if $M_j \notin \Omega_\epsilon(M_i)$ and $M_j \in \Omega_g(M_i)$:

$$P_{ij} = \sum_{M' \in \Omega(M_i)} \left[e^{\mathbb{A}(M_i).\delta}\right]_{MM'} .P(M', M_j)$$

– if $M_j \in \Omega_\epsilon(M_i)$ and $M_j \in \Omega_g(M_i)$:

$$P_{ij} = \left[e^{\mathbb{A}(M_i).\delta}\right]_{ij} + \sum_{M' \in \Omega(M_i)} \left[e^{\mathbb{A}(M_i).\delta}\right]_{MM'} .P(M', M_j)$$

– if $M_j \notin \Omega_\epsilon(M_i)$ and $M_j \notin \Omega_g(M_i)$

$$P_{ij} = 0$$

These corollaries are deduced from theorems 12.1 and 12.2 using the relation $p_{ij} = Q_{ij}(\infty)$, and the expression of the mean sojourn time $\mathbb{E}(t) = \int_0^\infty \tau.dF_d(\tau)$.

12.4.5. *Application to a case study*

Let us consider a system being used to process two kinds of jobs, J_1 and J_2. The requests for these jobs in the system may be described by a Poisson process with a rate α (time between two arrivals is a random variable following an exponential law of rate α). Each job requesting the system may be of type J_1 with the probability p or of type J_2 with the probability $(1 - p)$. When the system is requested, it will process the job in a fixed time delay δ_1 for a job of J_1 type or δ_2 for a job of J_2 type. During the job execution, the system may fail, its failure being characterized by a constant failure rate λ. The system is then repaired and will take back the job from the state where it was interrupted by the failure occurrence. The repair time follows a normal distribution law of average μ and variance σ^2. At the end of a job process, it must be removed from the system in order to process another job. This removal time has a random time exponentially distributed with the rate ε.

Figure 12.1 summarizes the behavior of this system.

Since the GSPN has a single token, there are as many markings as places. The reachability graph comprises seven nodes, the incidence matrix of this graph is the following:

$$G_{PN} = \begin{bmatrix} 0 & T_0(ex) & 0 & 0 & 0 & 0 & 0 \\ 0 & 0 & T_1(im) & T_1^{'}(im) & 0 & 0 & 0 \\ 0 & 0 & 0 & 0 & T_3(ex) & 0 & T_2(de) \\ 0 & 0 & 0 & 0 & 0 & T_3^{'}(ex) & T_2^{'}(de) \\ 0 & 0 & T_4(gen) & 0 & 0 & 0 & 0 \\ 0 & 0 & 0 & T_4^{'}(gen) & 0 & 0 & 0 \\ T_5(ex) & 0 & 0 & 0 & 0 & 0 & 0 \end{bmatrix} \begin{matrix} M_0 = P_0 \\ M_1 = P_1 \\ M_2 = P_2 \\ M_3 = P_2^{'} \\ M_4 = P_3 \\ M_5 = P_3^{'} \\ M_6 = P_4 \end{matrix}$$

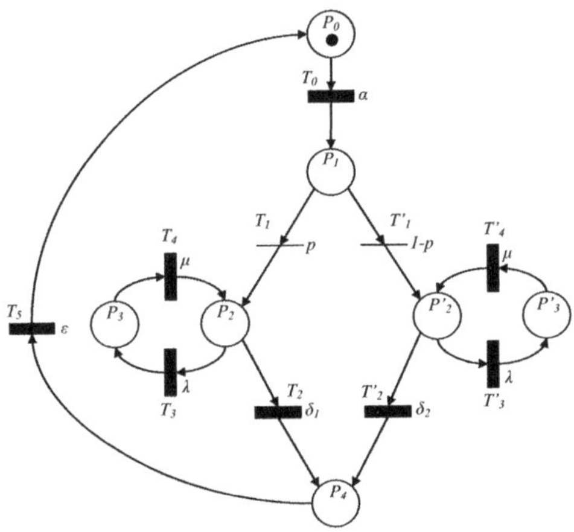

Figure 12.1. *Underlying PN of the model*

The reachability graph reduced to the tangible markings (marking M_1 is a vanishing marking wherein the system does not spend time) can be described by the following incidence matrix:

$$G_{PNR} = \begin{bmatrix} 0 & T_0(ex), T_1(im) & T_0(ex), T_1^{'}(im) & 0 & 0 & 0 \\ 0 & 0 & 0 & T_3(ex) & 0 & T_2(de) \\ 0 & 0 & 0 & 0 & T_3^{'}(ex) & T_2^{'}(de) \\ 0 & T_4(gen) & 0 & 0 & 0 & 0 \\ 0 & 0 & T_4^{'}(gen) & 0 & 0 & 0 \\ T_5(ex) & 0 & 0 & 0 & 0 & 0 \end{bmatrix} \begin{matrix} M_0 = P_0 \\ M_2 = P_2 \\ M_3 = P_2^{'} \\ M_4 = P_3 \\ M_5 = P_3^{'} \\ M_6 = P_4 \end{matrix}$$

Applying theorems 12.1 and 12.2, we obtain the following kernel of the semi-regenerative process:

$$Q(t) = \begin{bmatrix} 0 & p\left(1-e^{\alpha.t}\right) & (1-p)\left(1-e^{\alpha.t}\right) & 0 & 0 & 0 \\ 0 & 0 & 0 & \left\{\begin{array}{l} 1-e^{-\lambda.t}, t<\delta_1 \\ 1-e^{-\lambda.\delta_1}, t \geqslant \delta_1 \end{array}\right. & 0 & \left\{\begin{array}{l} 0, t<\delta_1 \\ e^{-\lambda.\delta_1}, t \geqslant \delta_1 \end{array}\right. \\ 0 & 0 & 0 & 0 & \left\{\begin{array}{l} 1-e^{-\lambda.t}, t<\delta_2 \\ 1-e^{-\lambda.\delta_2}, t \geqslant \delta_2 \end{array}\right. & \left\{\begin{array}{l} 0, t<\delta_2 \\ e^{-\lambda.\delta_2}, t \geqslant \delta_2 \end{array}\right. \\ 0 & F(t) & 0 & 0 & 0 & 0 \\ 0 & 0 & F(t) & 0 & 0 & 0 \\ 1-e^{\varepsilon.t} & 0 & 0 & 0 & 0 & 0 \end{bmatrix}$$

where $F(t) = \frac{1}{\sigma\sqrt{2\pi}} \int_{-\infty}^{t} \exp\left(\frac{-(t-\mu)^2}{2\sigma^2}\right) .dt$ is the distribution function of the repair time.

From this kernel, we obtain the transition probability matrix of the embedded Markov chain:

$$P = \begin{bmatrix} 0 & p & 1-p & 0 & 0 & 0 \\ 0 & 0 & 0 & 1-e^{-\lambda.\delta_1} & 0 & e^{-\lambda.\delta_1} \\ 0 & 0 & 0 & 0 & 1-e^{-\lambda.\delta_2} & e^{-\lambda.\delta_2} \\ 0 & 1 & 0 & 0 & 0 & 0 \\ 0 & 0 & 1 & 0 & 0 & 0 \\ 1 & 0 & 0 & 0 & 0 & 0 \end{bmatrix} \begin{matrix} M_0 = P_0 \\ M_2 = P_2 \\ M_3 = P_2' \\ M_4 = P_3 \\ M_5 = P_3' \\ M_6 = P_4 \end{matrix}$$

12.5. Synchronization problem in generalized stochastic Petri nets

12.5.1. *GSPN with internal synchronization*

A GSPN with internal synchronization [DAN 95] is an interesting way for a modular and functional modeling of a stochastic system. Each component and/or each functionality of the system is described by a stochastic PN and these different PNs are synchronized by the exchange of synchronization signals.

DEFINITION 12.4.– A generalized stochastic PN with internal synchronizations ($GSPNSyI$) is a set of generalized stochastic PN $R_{GSSyI} = \{R_{GS1}, R_{GS2}, ..., R_{GSr}\}$, such that $\forall i = 1, 2, ..., r$, R_{GSk} be a triplet $R_{GSk} = \langle R_{GSnk}, !S_k, ?S_k \rangle$ where:

- R_{GSnk} is a generalized stochastic PN;
- $!S_k$ is a set of synchronization signals emitted by R_{GSk};
- $?S_k$ is a set of synchronization signals received by R_{GSk}.

The synchronization signals are emitted by places when they receive at least one token which are received by transitions.

These PNs include four transition types distributed in the following four sets:

- T_{in} is the set of non-synchronized immediate transitions;
- T_{is} is the set of synchronized immediate transitions;
- T_{tn} is the set of non-synchronized timed transitions;

– T_{ts} is the set of synchronized timed transitions.

Evolution rules of a $GSPNSyI$:

– a synchronized transition is authorized if its waited synchronization signal is present;

– a non-synchronized transition is fireable if it is enabled;

– a synchronized transition is fireable if it is both enabled and authorized.

Graphically, the emission of a synchronization signal by a place (called emitter place) is represented by a broken arrow starting from the place. The reception of a synchronization signal, by a transition (called receiver transition), is represented by a broken arrow coming on the transition (Figure 12.2).

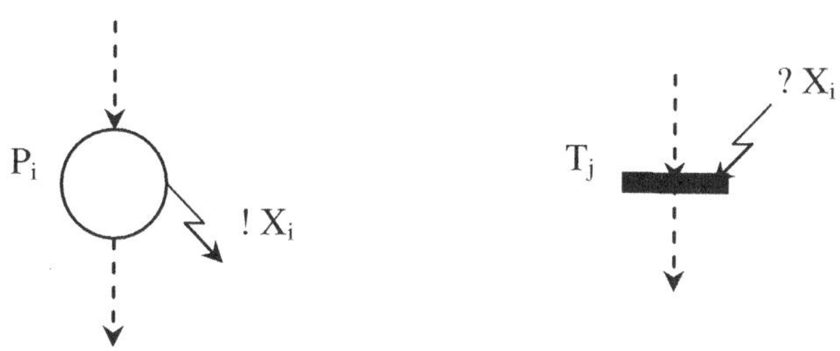

Figure 12.2. *Emission of a signal X_i by P_i and receipt of X_i by T_j*

The emitter places of an R_{SGi} send signals toward transitions of other R_{SGj}, meaning that the marking of the emitter places influences the evolution of the other PNs. A signal received by a transition may be a Boolean expression of one or more signals emitted by places of other PNs.

The use of stochastic generalized PNs with internal synchronizations for the modeling of a large system gives a greater readability to the obtained model, each component of the model having its own submodel independent of the submodels of other components.

The fact remains that the "signal" concept is theoretical and ambiguous. There is effectively no advice on how to implement a "signal". Does the signal have a duration? Could the signal be shared between multiple receivers? Is the signal managed as a message and if yes with what protocol type?

We will now tackle this problem in the following section.

12.5.2. *SPN with predicates and assertions*

In definition 12.4, the signal exchange seems to act as a semaphore but nothing is said about the existence continuity of the signal. Does the firing of the receiver transition reset the signal?

If the signal acts as an event, two cases may arise:

– the transition of the receiver is enabled and it may be fired by the event occurrence;

– the transition of the receiver is not enabled and the event is lost. The transition firing must wait for its enabling and a next event occurrence.

If the signal acts as a message, the receiver transition must be able to analyze its contents. It is then necessary to extend the definition of the used PNs in associating a predicate with this transition and the ability of the emitter place to produce a message content. Interpreted PNs are provided with such abilities that should also be attributed to stochastic PNs. In addition to these abilities, the transmission protocol of the message has to be defined. Is it a production–consumption protocol, a broadcasting protocol or something else? In the PN dedicated software, it is up to the user to find out the exact implementation of such mechanisms.

With CPN Tools software, the problem is solved using guards with the transitions. The guard condition acts as a Boolean condition which must be right to allow the firing of an enabled transition. This Boolean can be set by the emitter or be the result of an assertion evaluation.

This assertion can be seen as a function allowing us to take account of variables and the marking components (numbers and colors of tokens). It is then possible to easily implement the desired protocol.

We will now show how this may be achieved with MOCA RP software on the example of a set of three similar entities being repaired after failure. Here, only two repairmen are present and able to repair any of the entities.

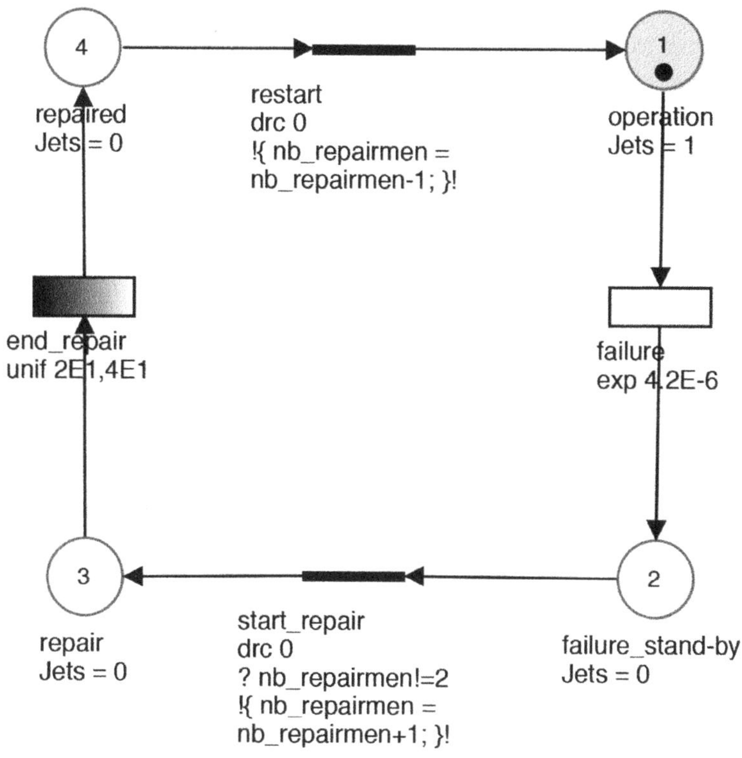

Figure 12.3. *PN representation of the first entity*

Figure 12.3 presents the SPN model of one entity. This model is the same for each of the other entities (the places are numbered similarly 7, 8, 9, 10 and 11, 12, 13, 14). The entity being initially in operational state (one token in the place number 1 also called "operation") remains here until its failure occurs. The transition "failure" is a stochastic transition characterized by an exponential law in the case presented in Figure 12.3, but any other probability law may be considered in order to describe the occurrence failure process (the behavior of the operating delays before failure). After firing this transition, the entity is in its state waiting for a repairman to be available (place 2 "failure_stand-by" marked). The number of available repairmen is given by the marking of the place 5 called "available_repairmen" of Figure 12.4. When a repairman is requested, the immediate transition "repairman_requested" is fired. The request of a repairman due to the occurrence of a failure of one entity is modeled by the guard of this transition "? $\#2 == 1 \mid \#8 == 1 \mid \#12 == 1$" representing the state when the place 2 (Figure 12.3) contains one token ($\#2 == 1$). The guard of this transition also represents further options of its firing (if one of two other entities fails) (places number 8 or 12 marked, in similar models of entities). This transition has priority (inscription "PRIO 1") in relation to the transition "start_repair" (in Figure 12.3), enabled in the same time. Also, for this transition, several firings are prevented at the same time (inscription "LIMIT") in order to avoid several repairmen being requested for a single failed entity. The number of repairmen engaged in an operation is not only given by the marking of the place "busy_repairmen" but also by the value of the variable "$nb\_repairmen$". If the value of this variable is different from the total number of available repairmen, modeled by the guard of transition "start_repair", this transition is fired and the entity is in its repair state (place "repair" in Figure 12.3). When the transition "start_repair" occurs, the variable "$nb\_repairmen$" is incremented by one (the inscription "$!nb\_repairmen = nb\_repairmen + 1;$"). Here, it is necessary to use the variable "$nb\_repairmen$" and not the actual marking of the place "busy_repairmen" in order to prevent the firing of the transition "start_repair" for a third failed entity, if the other two entities previously failed are still being repaired. The entity stays in its repair state (place 3) until its repair is done. The transition "end_repair" is a stochastic one characterized by a uniform law of repair delays in a given interval, but, such as for the "failure" transition, any other random law may be considered in order to describe the repair process. When the entity is repaired (place 4 called "repaired" is marked), first the transition "repairman_released" (Figure 12.4) is fired and then repairmen become available. This transition also has a guard "?#4 == 1 | #10 == 1 | #14 == 1" in order to represent this situation when a repair ends (the place 4 is marked for the entity represented in Figure 12.3 or for the other two entities of the system: places 10 and 14 marked). The behavior of this transition is similar to the behavior of the previously described transition "repairman_requested". Finally, the immediate transition "restart" (Figure 12.3) is fired in order to put the entity in its operating state. When this transition is firing, the value of variable "$nb\_repairmen$" is updated (inscription "!$nb\_repairmen = nb\_repairmen - 1$;").

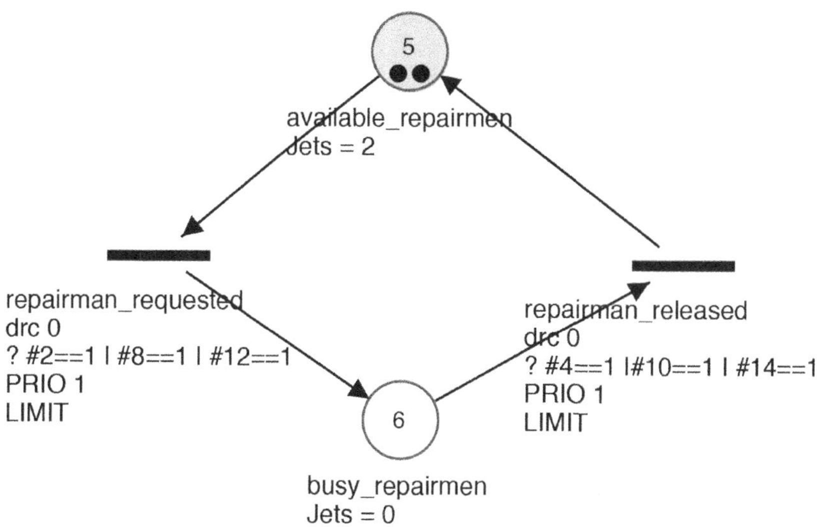

Figure 12.4. *PN representation of the repairmen*

We can remark here two types of synchronizations:

– Synchronization by event: an event is generated by one PN and it is used instantaneously to validate the firing of a previous enabled transition (by the corresponding marking) in other PNs. Afterwards, the event is "consumed" immediately, an event has no lifespan. This type of synchronization is modeled by the firings of two immediate transitions "repairman_requested" and "start_repair". When place 2 is enabled in the PN of entity, its marking is used as an event to activate the guard of the transition "repairman_requested". At the same time, this transition is fired followed by the firing of the transition "start_repair" that removes the marking of place 2. Thus, the marking of place 2 acts as an event.

– Synchronization by message: a message is generated by one PN and it is used to validate the firing of a previous enabled transition (by the corresponding marking) in other PNs; a message is not "consumed" immediately, it has a certain lifetime. This type of synchronization is modeled by the variable "$nb\_repairmen$" representing the fact that a repairman is busy from the start to the end of a repair operation.

12.6. Conclusion

Extensions of stochastic PNs give an interesting set of concepts to model systems with internal stochastic behaviors. All necessary hypotheses allowing an equivalence with a Markov or related model are closely defined. They give access to analytical solutions for the assessment of any dependability parameter. Outside these hypotheses, the same models may be extended for simulation purpose, even if synchronization between submodels is needed. In this case, we showed that particular attention must be paid to the implementation of synchronization mechanisms. Other examples of all these principles will be presented in Chapter 13.

PART 4

Applications of Stochastic Petri Nets to Assessment Problems in Industrial Systems

Introduction to Part 4

In this part, we will present a set of application examples of stochastic PNs (SPNs) and their extensions for the modeling and the probabilistic assessment of dependability properties. Most of these examples have been studied in the context of industrial cooperation. The first example comes back to the case study of dynamic reliability already proposed in the first book of this series. The following two examples are related to the electrical energy field. The last two examples are dedicated to the problem of the impact of failures in the functional performance of systems in railway transportation and networked control systems (NCSs). All of these studies were performed with different software tools that we will briefly present.

Software tools

The amount in software tools dedicated to Petri nets (PNs) handling is so vast that it would be vain to attemp to list them all. For this reason, we will only mention a handful because they either have been used in examples presented in this book or they seemed to significant us. We will not provide a detailed description of their characteristics and performances but we will only give the essential properties.

It must be noted that the different tools do not implement all of the theoretical concepts of SPNs. It is for example the case of the n-validation concept. In this case, a redundancy cannot be modeled by multiple tokens in a place but must be represented by different sub-PNs for example. Likewise, the colored PN (CPN) concept which can be very useful for the modeling of complex systems is not always available.

The French company SATODEV proposes the set of software tools GRIF-Workshop© graphical interface for reliability forecasting (GRIF) [SAT 15] dedicated to dependability studies. The PN module based on the Monte Carlo simulation engine MOCA-RP© developed for nearly 30 years by the French oil company ELF Aquitaine (now TOTAL) allows for the simulation of systems behavior described by PNs. It is not a freeware and is available on Windows and Linux platforms. One of its main advantage is the ability to describe a system in several PNs being possibility synchronized by "messages". Some of the examples presented in this book has been processed with MOCA-RP. It does not include the concept of token color.

Developed at the University of Turin, Italy (Università degli Studi di Torino), Great SPN 2.0 (Graphical editor and analyzer for timed and SPNs) [TOR 08] is a software package for the modeling, validation and performance evaluation of distributed systems using generalized stochastic Petri nets (GSPNs) and their colored extension: stochastic well-formed nets. It is a freeware available on Linux and Mac OS platforms. Its main characteristic is the modularity allowing separate programs to cooperate in the construction and analysis of PN models by sharing files. Different analysis modules can be run on different machines in a distributed computing environment. Furthermore, the graphical interface has been recently ported under OSF Motif achieving high portability under different hardware platforms.

Already described in section 4.6.3, CPN Tools [AIS 15] is primarily dedicated to colored Petri nets. Nevertheless, it incorporates functionalities for time and stochastic variables management as we saw in section 6.5.3. We will see in section 14.1 how it can be used as a GSPN model. It is a freeware available on Windows platform originally developed by CPN group at Aarhus University (Denmark) and now by the Architecture of Information Systems (AIS) group of the Eindhoven University of Technology (Netherlands).

Developed at the Institute of Machine Components of the University of Stuttgart (Germany), the reliability, availability, logistics and inventory simulation tool (REALIST) software implements a new formalism of extended colored stochastic Petri nets (ECSPN) [POZ 04a, POZ 04b] that is coupled with a reliability block diagram (RBD) model. The ECSPN includes colored tokens, stochastic timed transitions (with generally distributed firing times), component age information and queuing discipline. The integration of the age into the tokens allows us to consider a firing policy of transitions taking into account age influence, and arc expressions for age manipulation. Additionally, an operational cost concept is implemented in order to take into account economical aspects of system operation and maintenance. This cost is implemented in two manners: as cost places and as cost transitions. Available on the Windows platform, REALIST package allows the assessment of numerous probabilistic measures in the field of dependability and economical fields by means of Monte-Carlo simulation. REALIST software is an interesting tool with numerous applications in several industrial fields such as: automotive and mechanical engineering, power engineering, production processes, etc. (see for example [ZEI 14]).

13

Application in Dynamic Reliability

The case study presented here comes from [SCH 06]. Even if this example does not reflect the complexity of large industrial systems, it has been intentionally chosen for its didactic purpose and because it has been widely used as a kind of benchmark in the dynamic reliability community (see [CHA 99, CHA 98, DUT 97, MAR 96, KER 00, ZHA 08] and section 5.8 of the first book of this series [AUB 15]).

The main interest of this example is to highlight the problem of modeling hybrid dynamical systems for dependability assessment. We will particularly study the influence of the functional behavior on the occurrence and nature of the failure modes and, reciprocally, the impact on the operating mode of failure occurrences in the control system.

13.1. Presentation of the system and hypothesis

The system is composed of a tank where the fluid level (h) must be kept between two values ($h_0 - 1$) and ($h_0 + 1$). The tank is supplied by a main pump $PO1$, an emergency pump $PO2$, and is drained by a valve V. A control system regulates the fluid level by switching on/off the pumps and the valve according to the fluid level measured by sensors (Figure 13.1).

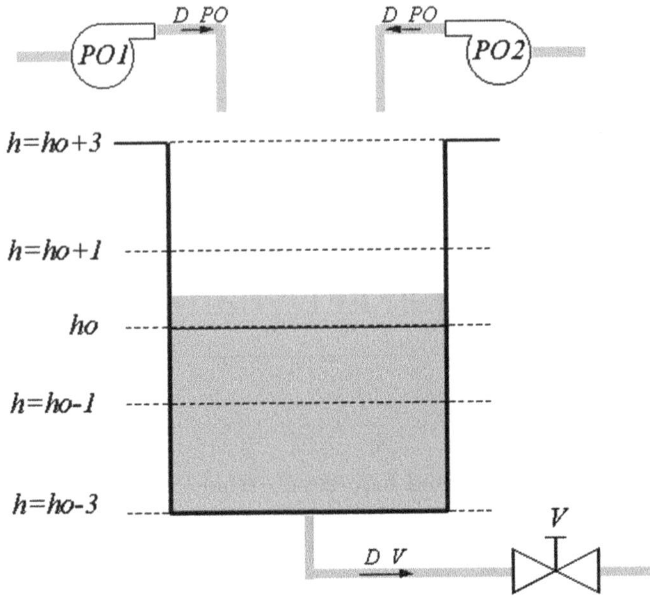

Figure 13.1. *The "tank-valve-pumps" system*

We make the following functional and dysfunctional hypothesis:

Functional hypothesis:

– the flows from the valve and the pumps are supposed to be identical and equal to $1.5m^3.h^{-1}$. The tank section is $1m^2$, so we can represent the flow as an hourly rate of level variation equal to $1.5m.h^{-1}$;

– initially (at time $t = 0$), the fluid level is supposed to be equal to $h_o = 3m$, the pump $PO1$ is on, the valve V is on, and the pump $PO2$ is off;

– sensors are located at different levels of the tank and, in nominal mode, the detection of a specific fluid level threshold will send orders to the actuators according to Table 13.1.

Fluid level h	Pump 1	Pump 2	Valve
$h < h_0 - \Delta h$	ON	ON	$Closed$
$h_0 - \Delta h \leqslant h \leqslant h_0 + \Delta h$	ON	OFF	$Open$
$h > h_0 + \Delta h$	OFF	OFF	$Open$

Table 13.1. *Control of actuators*

Dysfunctional hypothesis:

– we only consider failures of the valve and the pumps;

– these actuators are independent from each other and not repairable;

– we consider the following failure modes: the pumps can be stuck either in the present state (ON or OFF) or at the state ON, or at the state OFF. The valve can only be stuck at the present state;

– failure occurrence of these components obeys the exponential law with a rate $\lambda_1 = 2,2831.10^{-3}h^{-1}$ for the pump $PO1$, $\lambda_2 = 2,8571.10^{-3}h^{-1}$ for the pump $PO2$ and $\lambda_3 = 1,5625.10^{-3}h^{-1}$ for the valve V.

13.2. System modeling with Petri net

We now give the model of the level control system of the level in the tank, using the formalism of the "control interpreted Petri nets" [JAM 01] according to the rules of Table 13.1. This continuous variable h (the fluid level) is measured by sensors, and the information is used by the control system. The behavior of the controlled process (the fluid level in the tank) must be formalized too, as well as the interactions with the discrete part. The continuous part is modeled by PN on the basis of a sampling of the continuous variable h, which is symbolized by a set of tokens in a place. For example, we can assume that an elementary variation of liquid $\Delta h = \pm 10cm$ is modeled by the disappearance of a token (if $\Delta h = -10cm$) or appearance (if $\Delta h = +10cm$) in a specific place associated with the tank. The flow of the valve and the pumps are respectively symbolized by the disappearance and appearance of a token every 4 min.

Then, we build up several PNs representing the tank state (variation of the liquid level), the actuator state (on/off), the control system, and the communications by messages. Figure 13.2 shows these PNs and Tables 13.2 and 13.3 define the different places and messages.

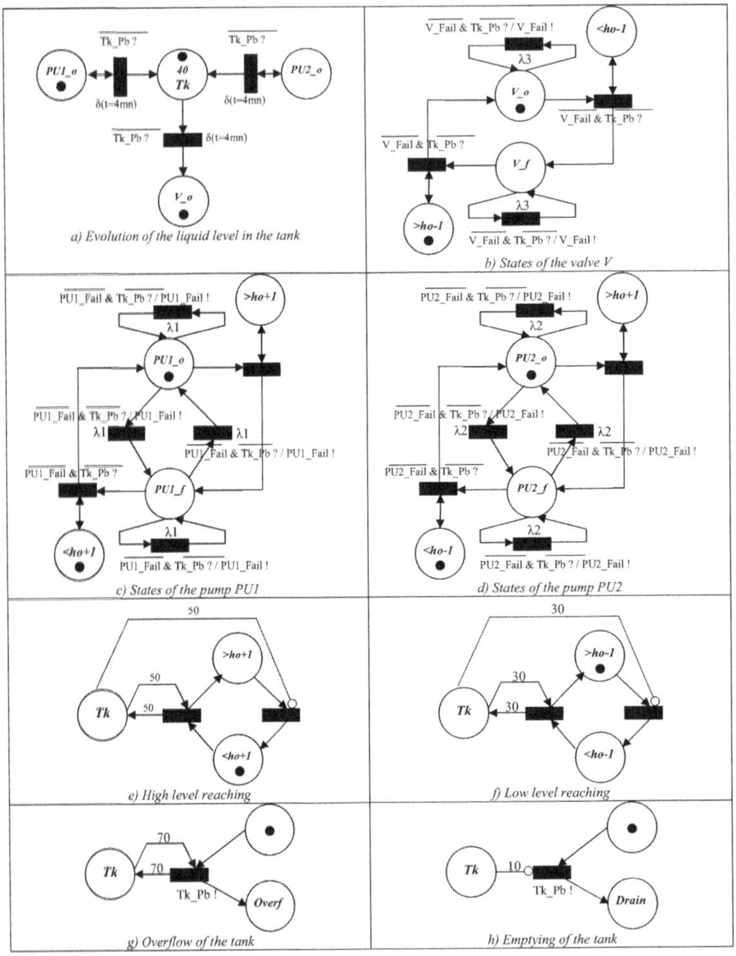

a) Evolution of the liquid level in the tank; *b) States of the valve V*; *c) States of the pump PU1*; *d) States of the pump PU2*; *e) High level reaching*; *f) Low level reaching*; *g) Overflow of the tank*; *h) Emptying of the tank*

Figure 13.2. *Modeling of the system in Petri nets*

The PN in Figure 13.2(a) formalizes the process evolution (level liquid in the tank). The PNs in Figures 13.2(b), 13.2(c) and 13.2(d) represent the states of the valve and the two pumps in accordance with the liquid level in the tank. These three PNs are conservative (p-semi flows). Formally, we can write: $M(V\_f) + M(V\_o) = 1$ $M(PU1\_f) + M(PU1\_o) = 1$ and $M(PU2\_f) + M(PU2\_o) = 1$. This means that each actuator is either ON or OFF. Eventually, the marking of the places in Figures 13.2(e), 13.2(f) and 13.2(g) allow us to determine the liquid level and to observe the feared events (overflow and emptying). These nets model sensor information. Figures 13.2(b) to 13.2(g) contain the control part, and Figure 13.2(a) contains the process.

Place	System state
T_k	Tank level (token number)
PU1_o	Pump 1 is ON
PU1_f	Pump 1 is OFF
PU2_o	Pump 2 is ON
PU2_f	Pump 2 is OFF
V_o	Valve is ON
V_f	Valve 1 is OFF
$<$ ho-1	Liquid level if lower than ho-1
$>$ ho-1	Liquid level is superior than ho-1
$<$ ho+1	Liquid level if lower than ho+1
$>$ ho+1	Liquid level is superior than ho+1
Overf	Tank overflow
Drain	Tank emptying

Table 13.2. *Description of the places*

Message (Boolean)	System state	Initial value
Tk_Pb	Overflow or emptying of the tank	FALSE
V_Fail	Valve failure	FALSE
PU1_Fail	Pump 1 failure	FALSE
PU2_Fail	Pump 2 failure	FALSE

Table 13.3. *Description of the messages*

The tool MOCA-RP© (see Introduction to Part 4) was used for the modeling of the system.

13.3. Methodology application

At the design stage of control systems, a verification phase to check the "good" properties of the functions must be performed before any quantitative dependability study. In case of PNs, this phase consists of verifying the liveness and the safeness of all the sub-nets modeling the functions. For more details concerning the study of properties, we can refer to [SCH 04].

At the end of the qualitative analysis, several feared events are identified. The aim of this study is to evaluate the occurrence probability of these events. The control system is assumed not to be repairable. Thereby, after quite a long time, the system reaches a steady state after the occurrence of one or more failures. We identified three sets of steady states:

– set 1: states corresponding to the overflow of the tank;

– set 2: states corresponding to the emptying of the tank;

– set 3: all the remaining steady states. In every one of these states, the system no longer changes. The liquid level remains constant.

Specific places are added on the PN to detect the occurrence of the feared events (Figures 13.2(g) and 13.2(h)). These places are essential for quantitative analysis by checking the date of their marking during the simulation.

13.4. Construction of an aggregated Markov graph

There are several conceivable means to build up an aggregated graph: the first consists of displaying the reachability graph to identify all the elementary states, and then lumping them together in macro-states (set of elementary states) in accordance with a common operating mode. This option leads to a combinatorial expansion in case of complex systems. In this example, the system contains at least 700 states after sampling the variable h.

Another technique consists of displaying an *a priori* reduced graph without looking for the elementary states. The identified "macro-states" are the following:

– Nominal mode: this macro-state is made up of all the elementary states in the absence of failure.

– $ER1$: all the states of the system in the presence of the feared event "overflow" are included in this set. Some combinations of actuator failures can lead to this set of steady states, for example when the valve is stuck at "off" and the two pumps are "on".

– $ER2$: idem as above for the feared event "emptying".

– $ER3$: all the elementary states defining a variation of h equal to zero.

– Degraded modes: we borrow this terminology to refer to the transient macro-states whose elementary states do not belong to one of the previous state classes. In fact, these state sets are characterized by the presence of one or several failures, but a minimal operation is provided by the system. For example, the presence of the failure "Pump $PO1$ stuck at position Off" does not involve a dramatic consequence for the control system which controls the level using the emergency pump $PO2$. Finally, we can find 15 degraded modes.

It is important to remember that the passage from a functional or dysfunctional mode to another mode happens on the occurrence of one or several failures. Then, each macro-state is defined by the presence within the system of zero (nominal mode), one or several failures (degraded modes and steady macro-states). Each operating mode and the sets of feared events are featured by a combination of failures. In our case, it is easy to identify the combinations of failures specific to degraded modes (when the control is maintained), and to the sets of feared events.

Thus, the aggregated graph includes 19 macro-states, a part of them being shown in Figure 13.3. It is easy to check by combinatorial logic that the union of the macro-states appearing on the graph forms a complete set. For better readability, not all arcs and states are shown.

The following convention will be used for a better understanding of Figure 13.3:

$P1$: Pump $PO1$ stuck at ON, : $\overline{P1}$ Pump $PO1$ stuck at OFF,

$P2$: Pump $PO2$ stuck at ON, : $\overline{P2}$ Pump $PO2$ stuck at OFF,

V : Valve V stuck at ON, : $\overline{V}$ Valve V stuck at OFF.

The graph (Figure 13.3) shows the event combinations leading to the feared states (ER1, ER2 and ER3). Let us remark that methods like fault trees give a statistical representation of these combinations, and thus provide pessimistic results for the reliability of the system. Using a Markov graph, we avoid such disadvantages because the dynamic feature is considered due to the variables p_i. These coefficients weigh the failure rates associated with the arcs and are dependent on the sojourn time in the upstream states of the arcs. Furthermore, this graph shows all the sequences leading the system from the nominal mode to a feared state.

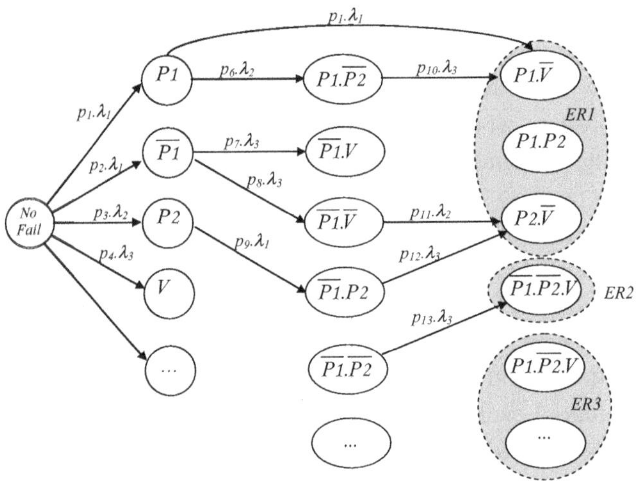

Figure 13.3. *The aggregated Markov graph of the system*

The performances of a Monte-Carlo simulation are dependent on the number of stories N to be simulated, and the duration of each story T. The simulations are performed with N=1000. This value is more than enough to obtain accurate results for the system. The duration of stories is fixed to T=1000 hours. This value allows us to obtain the final stationary distribution for all the evaluated variables by simulation.

A direct simulation of the behavioral modeling, including the failures, would have required a much longer simulation time to observe the occurrence of a feared event.

Some probabilities p_i can sometimes be deduced from a logical reasoning. Let us consider for example that the coefficient p_4 is equal to zero. We just have to carry out the following deduction: the failure mode of the valve is "stuck at the actual position". However the macro-state origin of the arc "$p_4.\lambda_3$" relates to the nominal mode ("No Fail"), and in this mode, the valve is never OFF according to the functional hypothesis, i.e. the valve and the pump 1 are ON, the pump 2 is OFF. Thus, the conditional probability of the valve being OFF, given that the system is in the state "No Fail", is equal to zero.

The same reasoning cannot be applied to p_7 and p_8. In the degraded mode named $\overline{P1}$, the valve is alternately ON and OFF. When a failure of the valve occurs at time t_f, its failure mode will depend on the state of the valve at t_f. In probability terms, the privileged failure mode depends on the sojourn time in the states "valve ON" and "valve OFF". These times depend on the process dynamic and can easily be evaluated by simulation. The aim is to deduce asymptotic probabilities for the valve being OFF and ON, given that the pump is stuck at OFF ($\overline{P1}$).

The tool MOCA-PR allows us to deduce these coefficients by a Monte-Carlo simulation. So, to assess the probabilities p_7 and p_8, the system is configured "Pump PO1 stuck at OFF" by artificially injecting the suitable failure, and assuming no more failures should affect the system afterward. The valve can only have two different states, and thus, we can write the relation: $p_7 + p_8 = 1$.

This constraint is formalized on the PN with the invariant (P-semiflow): $M(V\_o) + M(V\_f) = 1$. In other words, the state of the valve is either "ON", or "OFF". This relation is also true for the mean marking. Then, we have $p_7 = M^{\infty}(V\_o)$ and $p_8 = M^{\infty}(V\_f)$, i.e. the asymptotical probability is equal to the mean marking. So, this method allows us to assess all the coefficients by simulation.

Then, we just have to convert the Markov graph into simultaneous differential equations (Chapman–Kolmogorov) and solve them using classical techniques.

Figure 13.4 shows the probability evolution of being in one of the three possible feared states ER at time t for the considered system.

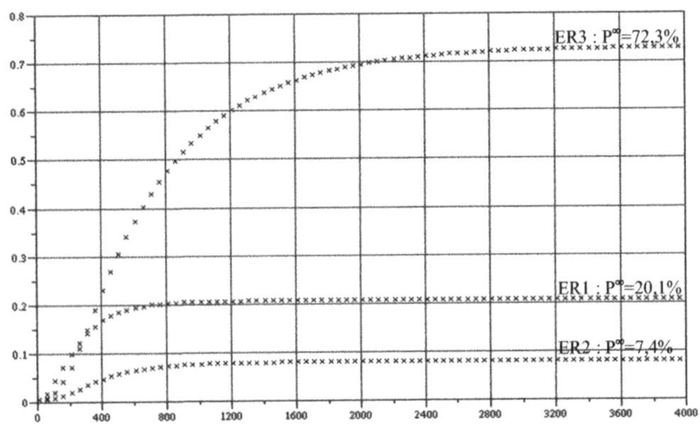

Figure 13.4. *Probabilitly evolution of the feared event ER*

These curves show that after quite a long time, the probability of being in the ER3 state is the highest. Overflowing and emptying only relate to 27% of the system failure. Afterward, the designer may look for some corrective modifications on the control system to decrease these probabilities and improve dependability if safety requirements are not satisfied (by adding fault tolerance device for example). Note that these results are nearly identical to the direct simulation of the system, but it needs a longer simulation time. For this example, the direct simulation of the global model was performed with various solver types using Mathlab Simulink Stateflow. The most significant result needed a computation time of about 70 h (10,000 stories). To get the equivalent Markov graph, the global time for the analysis may be estimated to be 5 h and the simulation for determination of weighting coefficients needed 1 h. The computation time is then highly reduced and in addition, the model (Markov graph) includes all the failure sequences leading to feared events.

13.5. Conclusion

The interest of this approach is to combine two different resolution techniques. On the one hand, an analytical method was developed by building up an aggregated Markov graph. It allows the designers to have an exact representation and a better interpretation and overview of the states, although the combinatorial explosion problems induced by this kind of method are avoided. On the other hand, we used Monte-Carlo simulation of PN submodels to evaluate weighting coefficients of failure rates.

14

Classical Dependability Assessment

14.1. Availability study of a nuclear power plant subsystem

This example is an extract from a real and more complex case study, developed by Electricité de France (EDF) for the Approdyn project [AUB 12]. The modeled system is a controlled system composed of one physical subsystem and one control subsystem. The physical subsystem is composed of two feed-water turbo-pumps (TPA) working in parallel. Each pump is composed of two subsystems: a turbine part (denoted as T) and an out-of-turbine part (denoted as Out-of-T). If one of these subsystem fails, the corresponding feed-water pump fails. The reliability block diagram of this physical system is given in Figure 14.1.

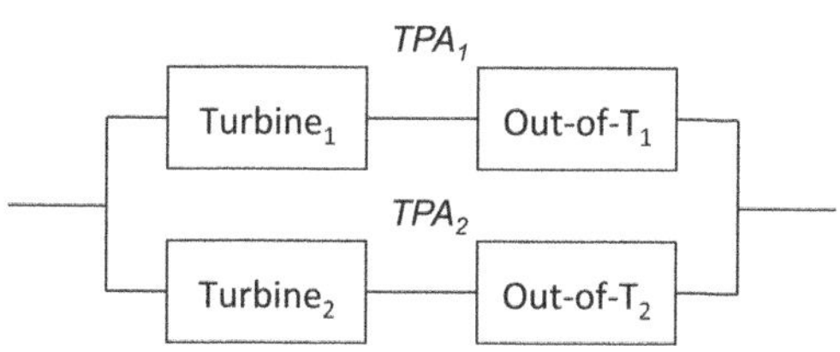

Figure 14.1. *Reliability block diagram of the TPAs system*

The data characterizing the failure and reparation process of each component are presented in Table 14.1.

TPA	$MTTF_T$	$MTTF_{HT}$	$MTTR_T$	$MTTR_{HT}$
TPA_1	6780	6854	4	48
TPA_2	2260	6.8×10^6	48	288

Table 14.1. *Components MTTF and MTTR, (in hour)*

The failure phenomena are characterized by the exponential law, with the following cumulative distribution function:

$$F(t) = 1 - e^{-\lambda_c.t}$$

where: $\lambda_c = 1/MTTF_c$ ($c = T$ or out-of-T component) is the rate parameter.

For reparation times an Erlang law is considered. Its cumulative distribution function is the following:

$$F(t) = 1 - \sum_{k=0}^{n-1} \frac{1}{k!}.e^{-\mu_c.t}.(\mu_c.t)^k$$

where $\mu_c = 1/MTTR_c$ ($c = T$ or out-of-T component) is the rate parameter and $n = 2$ is the order parameter.

The control subsystem describes the specifications of the control used for this system, which is the following. If both pumps are in the ON state, the system is working at nominal parameters. If one of the components of a pump fails, the other component of the same pump is stopped and a reparation order is given. The system works in a degraded operating mode. When the repair is finished, the system restarts immediately; the repaired pump and the system are working again at nominal parameters. When both pumps are in a failure state, the entire system fails.

14.1.1. *CPN modeling*

14.1.1.1. *GSPN with CPN tools*

The CPN model of this system (a physical part and a control part) is implemented using the CPN tool freeware.

To assess dependability indicators, the stochastic events, such as failures and repairs which occur in the physical part of the system, must be taken into account. They are modeled by stochastic transitions fired after a random enabling time. At the same time, deterministic reaction of the control part must be modeled using "immediate transitions" that occur instantaneously in time.

These requirements are covered by the class of GSPNs defined in [AJM 84]. Immediate transitions have priority over the stochastic transitions: if in a given marking, immediate and stochastic transitions are simultaneously enabled, the immediate transitions are fired first in zero time once they are enabled. However, when a token enables several stochastic transitions (competition between transitions T_1 and T_2 in Figure 14.2), it is assigned to the transition for which the realization of a random time variable (characterized by a transition rate λ) is smaller. This transition wins the token but is only fired after the token sojourns in the input place according to random time variable.

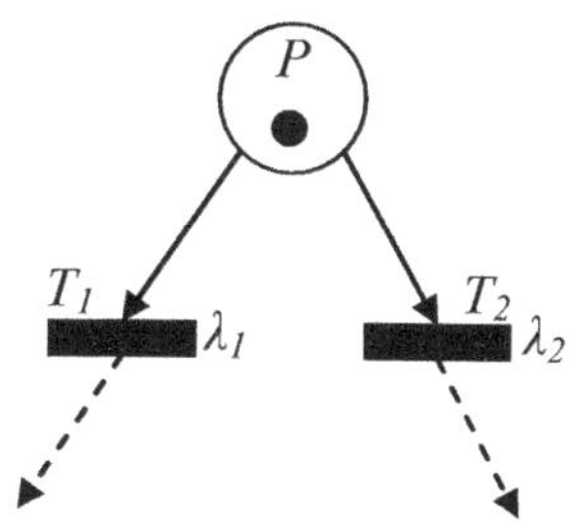

Figure 14.2. *Concurrence of GSPN stochastic transitions*

In a timed CPN [JEN 09], the time is given by a global clock. In addition to their color, the tokens in a marking contain time stamps. When a transition is enabled, it is fired and the time stamps of tokens deposited in its output places are changed. In these places, the tokens remain "frozen" and cannot be used to enable other transitions as long as the current time of the model (as given by the global clock) remains smaller than their own time stamps. As soon as the current time model becomes equal to or greater than the time stamp of the tokens, the latter can enable other transitions which are instantly fired. In other words, the time stamp describes the "earliest" model time from which a token can be used. Consequently, this behavior matches the formalized theoretical behavior of P-timed PN operating at its maximum speed.

Regarding the immediate transitions, their behavior is the same in the GSPN and CPN models, but, unfortunately, the behavior of the stochastic transitions is quite different in a timed CPN model compared to a GSPN formal model. If the PN of Figure 14.2 is a timed a CPN, as soon as the token enters place P, it is assigned to one of the two transitions, T_1 or T_2, and this transition is immediately fired. The token does not sojourn in place P and its time stamp is modified according to the rate of the corresponding transition. Then the token sojourns in the output place as long as required by its time stamp.

To solve the problem of stochastic transition competition and to force a sojourn time in the input place of a stochastic transition, the idea is to deal with the competition stochastic choice before enabling T_1 and T_2 transitions. This anticipated process is done by modeling the GSPN of Figure 14.2 by the CPN of Figure 14.3:

– Transitions "Fault 1" and "Fault 2" correspond to the transitions T_1 and T_2.

– Place "Working" corresponds to place P.

– Additional transition "Starting" takes a random value from distributions associated with T_1 and T_2 and characterized by λ_1 and λ_2 rates; the smallest of these random values is the expected sojourn time in place P. It is allocated to time stamp of the token deposited in place "Working"; this process is coded in ML language [HAR 00] and requires the definition of a more complex color containing the values of the two random variables in addition to token information.

– As soon as the global clock reaches a token time stamp, only the transition corresponding to the smallest value of random time variables (as defined in transition "Starting") must be fired; this can be done using a guard given to arc from place "Working" to transitions "Fault1" and "Fault2".

14.1.1.2. *The model of the subsystem*

The proposed model (Figure 14.3) ensures a behavior that is compliant with formal GSPN with "enabled memory" policy for firing transitions. The reasoning is generic and can be applied each time a competition between stochastic transitions occurs.

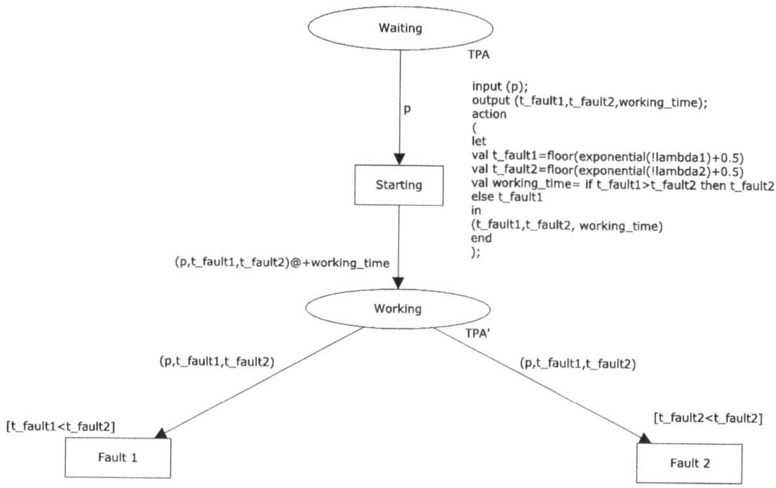

Figure 14.3. *GSPN modeling behavior with a timed CPN*

The system model has two levels of hierarchy: one for the components themselves and another for the coupling process and control subsystems. The hierarchy module is presented in Figure 14.4(a). This module links two models: the stochastic model representing the physical subsystem with failure and reparation probability distribution functions and the deterministic logical model which is a control subsystem. The top model contains the initial conditions and a place shared by the two modules, allowing us to check the system status.

The substitution transition "System" in Figure 14.4(a) is the physical TPA model represented in Figure 14.4(b).

The TPA model describes a generic TPA with two types of failures, illustrated in the RBD of Figure 14.1. Two tokens represent the TPA_1

and TPA_2. When the "Starting TPAs" transition is fired, the instants time of the next failures of the turbine and the out-of-turbine part of each TPA is calculated. These instants times are then used to determine the sojourn time of the token in the "TPAs in work" place. When the first failure occurs on "pump1" or "pump2", the corresponding transition is fired ("Fault Turbine" for the turbine part or "Fault HT" for the out-of-turbine part). The firings of these transitions change the status of the corresponding pump, from on ("true" color) to off ("false" color), that is represented by the color of the tokens in the "Status TPA" place. This place is associated with the shared place "Status of system" in the top module. The repair of the failed component is completed and when this reparation is finished ("Turbine repaired" or "HT repaired" transition) the pump returns in the "TPAs waiting" place and the status of the pumps is updated in the "Status TPA" place.

The control model, illustrated in Figure 14.4(c), corresponds to the substitution transition "Specification" of Figure 14.4(a). The control model is used to investigate the behavior of the physical system, to stop a TPA and to allow a reparation. The "System OK" place is associated with the the shared place "Status of system" in the top module. When the status of one pump changes from On to Off by the physical system, the "FTPA1" or "FTPA2" transition is fired and the pump is in the fail state ("TPA1 down" or "TPA2 down" places). In these states, if the components are repaired first, the status is updated to On by the physical system in the place "System OK". Now the transition "RTPA1" or "RTPA1" is fired to update the state of the control system. If one pump fails ("TPA1 down" or "TPA2 down" places) and the second pump fails "1'(pump1,false)++1'(pump2,false)", the "System failure" transition is fired and the system breaks down ("System KO" place).

14.1.2. *Reliability and dependability assessment*

The reliability and dependability assessment is provided by means of Monte-Carlo simulations, which is the only way to carry out the performance evaluation when the Markov hypothesis is not verified, due to Erlang laws modeling the repair processes. The following indicators are assessed: the availability, MTTFF, MTBF and MTTR for the entire system and for each pump (a subsystem composed of two components).

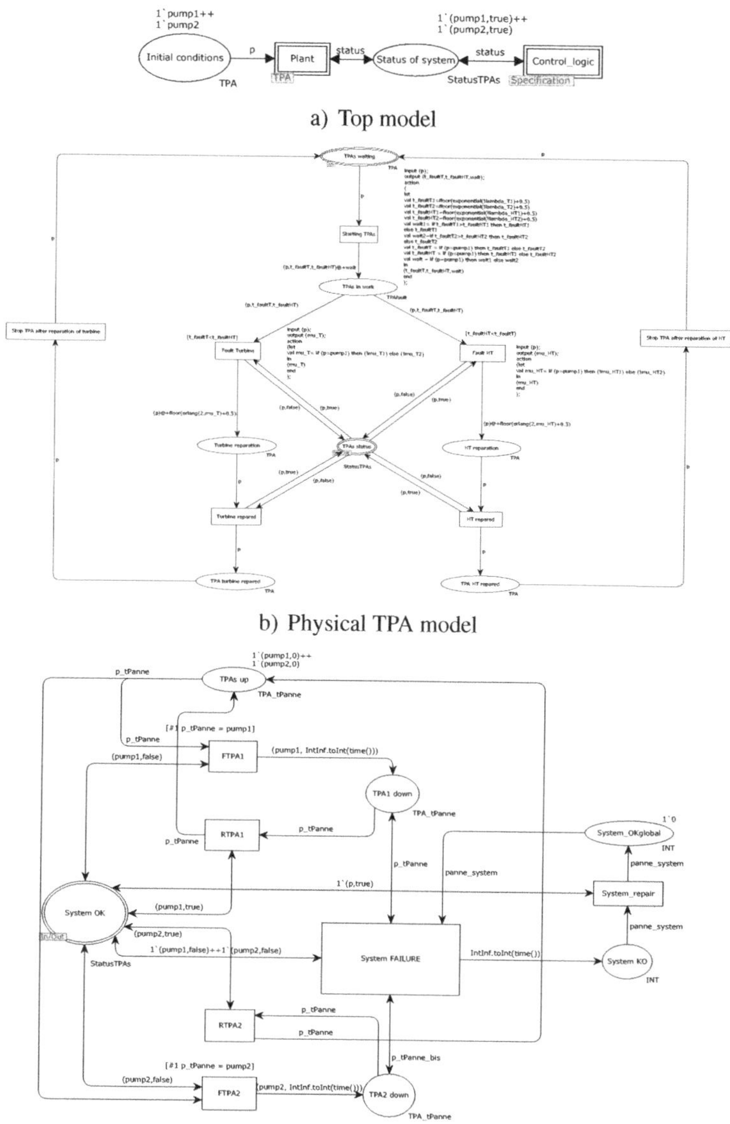

a) Top model

b) Physical TPA model

c) Logical control model

Figure 14.4. *CPN models associated with the case study*

The probabilistic indicators, such as availability, are measured by a probability and can be determined based on marking invariants (according to section 1.3.8, a marking invariant is a subset of PN places where the number of tokens is constant). A token, associated with a component, a subsystem or a system, evolves in places describing its state (waiting, operating, failing, etc.) and these places constitute a marking invariant. The probability, representing the indicator that should be estimated, is given by the ratio between the average marking of the place(s) that describe the state(s) characterizing the indicator sought and the sum of the average marking of all places belonging to the invariant, i.e. the number of tokens contained by the places subset:

$$P(state_I) = \frac{\overline{M}(state_I)}{\sum\limits_{P_i \in P_{subset_I}} \overline{M}(P_i)} \qquad [14.1]$$

where $state_I$ is the state characterizing the probabilistic indicator I, $\overline{M}(state_I)$ is its average marking and P_{subset_I} is the places subset of the invariant.

For example, for one system (number of tokens is equal to 1), its unavailability $\overline{A}$ can be estimated by the following equation:

$$\overline{A} = \Pr(state_{\overline{A}}) = \frac{\overline{M}(state_{\overline{A}})}{\sum\limits_{P_i \in P_{subset_{\overline{A}}}} \overline{M}(P_i)} = \overline{M}(state_{\overline{A}})$$

where $state_{\overline{A}}$ represents all the down states of the system.

Consequently, such probabilistic indicators can be estimated by average marking of the corresponding place(s).

The mean time indicators are measured by the average value of the sojourn delay in the place(s) characterizing the indicator sought.

$$MTI = \sum_{P_i \in P_I} \overline{D}(P_i) \qquad [14.2]$$

where MTI is the mean time of indicator I, P_i is a state characterizing the indicator I, P_I the subset of all these places and $\overline{D}(P_i)$ is the average value of the sojourn time in the place P_i given by Little's formula [10.4].

In the CPN tools, these indicators are supported by monitor functions developed in ML language used to inspect the CPN during its simulation. In our case, four types of monitors are used for the following indicators:

– MTTFF, for which the monitors should record only the time of the first entity failure;

– MTTR, for which the monitors should record every reparation time;

– MTBF, for which the monitors should record every failure time;

– unavailability, for which the monitors should record the duration of state when the entity is broken down.

These monitors are implemented for the two TPAs and for the whole system.

The statistical results obtained in the Monte-Carlo simulation (over 10,000 replications) are given in Table 14.2 where $95\%CI$ represents the half-length of a 95% confidence interval.

Indicators	Type	TPA1	TPA2	System
MTTFF	Average	3111.7	2234.9	6340.5
(in hour)	$95\%CI$	61.75	43.9	375
MTBF	Average	3144.2	2259.2	6312.8
(in hour)	$95\%CI$	62.4	44.35	373.35
MTTR	Average	26.4	48	16.4
(in hour)	$95\%CI$	0.5239	0.9428	0.9734
Unvailability	Average	0.00694	0.01901	0.00014
(%)	$95\%CI$	0.000113	0.000196	0.000012

Table 14.2. *System performance results*

The data obtained by Monte-Carlo simulation is also used to determine the empirical distributions of different mean times, which are given in Figure 14.5. As shown in Figure 14.5(a), the MTTFF of the controlled system is uniformly distributed, due to the mixture of several exponential and non-exponential laws. The distributions of MTTF and MTBF are quite similar in average and type of law. Indeed, the reparation rate is too small in comparison to the failure rate to modify the probability density function heavily. The empirical distribution of MTTR, Figure 14.5(c), is similar to an exponential distribution. This is due to the small values of reparation times (the probability of another event during the repair is relatively small) and because the repair starts instantaneously after a failure occurs.

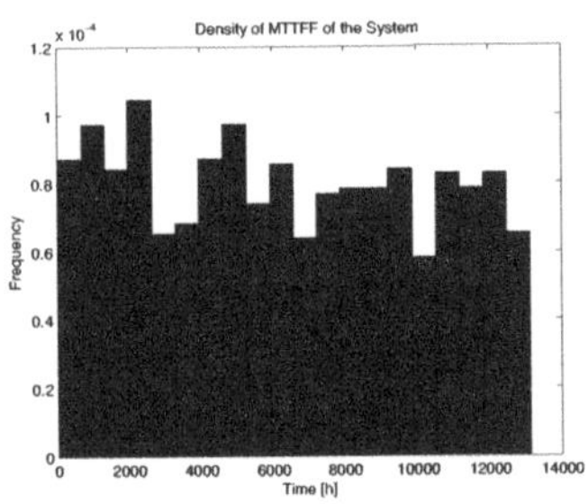

a) Mean Time To First Failure

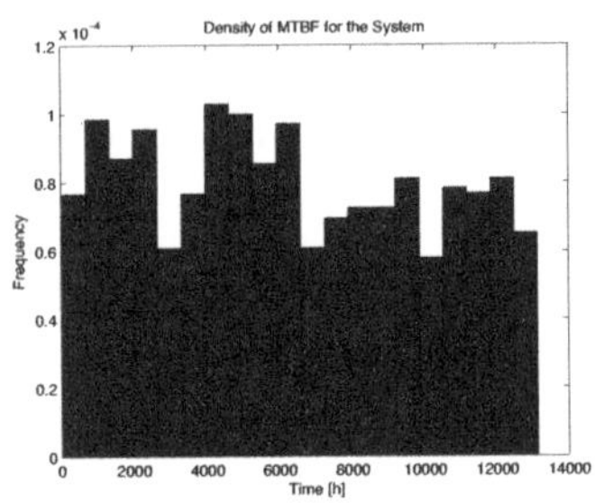

b) Mean Time Between Failure

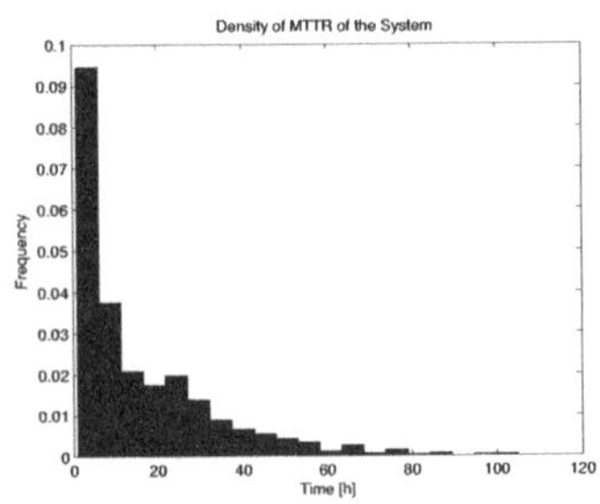

c) Mean Time To Repair

Figure 14.5. *Empirical distribution of the MTTFF, MTBF and MMTR of the whole controlled system*

14.1.3. *Conclusion*

In this section, an approach based on CPN has been proposed to cover the probabilistic dependability analysis. The probabilistic assessment is realized by means of Monte-Carlo simulation, without any modification of the stochastic CPN model. The probabilistic indicators are defined in terms of marking of different places of CPN and can easily be implemented by the monitor functions in the CPN tools.

14.2. Common causes failures in nuclear plants (safety oriented)

Highly redundant instrumentation and control (I&C) system in safety relevant industries are often used to satisfy requirements in dependability and safety. In these systems, homogeneous redundancies are generally preferred as their maintenance cost may be optimized. However, they are more sensitive to common cause failures and special attention should be paid to this problem. In this context, we present a study of the impact of common cause failures (CCF) on the dependability of such a system in the nuclear industry [BRI 14]. This system dedicated to safety functions is part of the in depth defence strategy of a nuclear power plant. The study is based on a colored PN model of the system in which the Atwood model of CCF is integrated.

14.2.1. *The Atwood model*

The β-factor model is the most widely used model for taking into account CCF within all types of systems, like nuclear power plants [USN 87] and, more generally, in the field of power systems [BRI 13]. It involves the failure of a a whole set of components when a common cause event occurs. This model is used when the system is composed of only a few components. However when the system is composed of dozens of identical or similar components, the assumption of failure of a whole set of components, when a CCF occurs, is very conservative (leads to an over estimation of the dependability).

The model introduced by Atwood [ATW 80] considers that the system components are subject to two types of failures: independent failures and shock. Two kinds of shock failures are defined: lethal shocks and partial (or non-lethal) shocks. In large redundant systems with N components, a shock is assumed to be non-lethal when it affects k components among N with $1 \leq k \leq N$. In the case of a non-lethal shock, only the failure of some components is considered. Each component then has a conditional probability p of failure. Individual failures, non-lethal and lethal shocks are assumed to follow independent processes. The occurrence frequencies of shocks (denoted as μ for non-lethal shocks and ω for lethal shocks) are assumed to be constant.

The failure rate of a specific component in a group of N elements, due to an independent failure or to a non-lethal shock is:

$$\lambda_1^N = \lambda_i + \mu \times p \times (1-p)^{N-1} \qquad [14.3]$$

The failure rate of a group of k components from N with $1 \leq k \leq N$ due to a non-lethal shock is:

$$\lambda_k^{(N)} = \mu \times p^k \times (1-p)^{N-k} \qquad [14.4]$$

The failure rate of N components due to a non-lethal and lethal shock is:

$$\lambda_N^N = \mu \times p^N + \omega \qquad [14.5]$$

For a specific component in a group of N components, the total failure rate is given by:

$$\lambda_{TOT} = \lambda_{IND} + \omega + \mu \times \sum_{k+1}^{N} C_{N-1}^{k-1} \times p^k \times (1-p)^{N-k} \qquad [14.6]$$

The Atwood model is considered representative for the phenomena leading to multiple failures in the large digital I&C systems. It allows us to represent CCF affecting only a part of all the components of the system. However, it introduces three parameters to estimate p, μ and ω. The results will be sensitive, of course, to the values used, usually given by expert opinion or by using default values. The default values commonly used for these parameters are the following:

$- \alpha = \frac{\mu}{\lambda_{TOT}} = 0.405$: the rate of non-lethal shocks;

$- p = 0.2$ or 0.33 or 0.5: conditional probability of component failure in a non-lethal shock;

$- \beta_{letal} = \frac{\omega}{\lambda_{TOT}} = 5.10^{-3}$: the rate of lethal shocks.

We propose to determine the values of these parameters by an approach based on the results of Monte-Carlo simulations.

14.2.2. *Case study*

14.2.2.1. *System architecture*

This protection I&C system contains four divisions, which are strictly identical, as represented in Figure 14.7. These divisions are physically separated. Each division is composed of five "Acquisition & Processing Units" (APU). The APU 0, 1 and 2 represent the subsystem A (SSA). The APU 3 and 4 represent the subsystem B (SSB). The gathering of APU into subsystems takes into account the allocation of control functions of the I&C system. For the same control function, there is one implementation in an APU of SSA and one implementation in an APU of SSB with different inputs and treatment and their outputs must be identical in the normal operating mode.

Different electronic boards C1 and C2 are included in each APU. Each APU contains one C1 board. The APU 0 and 1 contain four C2 boards, and the APU 3, 4 and 5 contain three C2 boards. These electronic boards are used for reception, processing and emission of signals.

A second type of partition is defined: the groups of APU (GAPU). A group of APU contains all of the APUi (i$\in [0, 4]$) of the four divisions.

14.2.2.2. *Modeling assumptions*

14.2.2.2.1. Assumptions for the electronic board modeling

A constant failure rate is considered for boards.

The C2 boards of an APU are considered as a series system.

The board failures can be detected by a set of self-tests. When a failure is detected by a self-test (SA), the detection time is considered as null. When a failure is not detected by a self-test (NSA), then it is detected offline during a periodic-test. For a given division, these periodical-tests take place every 18 months. So every quarter of this period, a division is tested during the periodic tests. After the periodic tests, the failed boards are repaired.

We assume that the detection of failures during a periodic-test does not cause an inspection on the other divisions, and that the whole system is operational during the periodical-test.

According to the supplier of electronic boards, their coverage rate of self-tests t_c is 100%, but this value seems to be ambitious. We add to the model the non-detected failures by self-test (NSA failures) in order to take into account the errors due to operation (different parameters or installation of boards from the nominal conditions specified by the supplier). Thus, the coverage rate is reduced by 85%. The global failure rate of the boards remains identical ($\lambda_{IND} = \lambda_{SA} + \lambda_{NSA}$). The rates of detected failures (λ_{SA}) and non-detected failures (λ_{NSA}) are adjusted by the following equations:

$$\lambda_{SA} = t_c \times \lambda_{IND}$$

and

$$\lambda_{NSA} = (1 - t_c) \times \lambda_{IND}.$$

14.2.2.2.2. Assumptions for global system modeling

The hazardous event is represented by the failure of the protection I&C system.

The occurrence of this hazardous event is based on the voting logic of the APU:

– an APU fails when a card C1 or C2 fails;

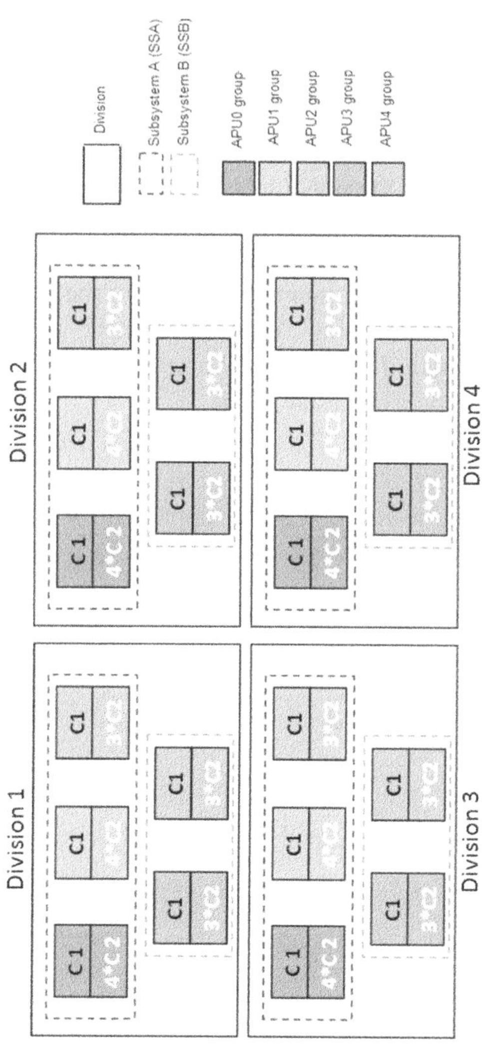

Figure 14.6. *Architecture of the case study I&C system for a nuclear power plant*

– a GAPU fails when 3 out of 4 APU fail (3oo4);

– a subsystem (SSA or SSB) fails when a GAPU is fails;

– the I&C system fails when a subsystem fails (1oo2) or when two subsystems are failed (2oo2) (two strategies of logic vote can be considered).

We assume that the mission time of the protection I&C system is 10 years. Indeed, this system is retrofitted only during the decennial maintenance operations of the nuclear plant. System unavailability can occur over 10 years. The system changes from the unavailable to the available state without being "as good as new" as it is not fully retrofitted in this period of 10 years. Some electronic boards may still fail when the system becomes available.

14.2.2.3. *CPN modeling*

We propose the use of a modular approach for system modeling. Thus, the CPN model (shown in Figure 14.7) is composed of the following modules:

– CCF generating;

– system representation;

– state system description.

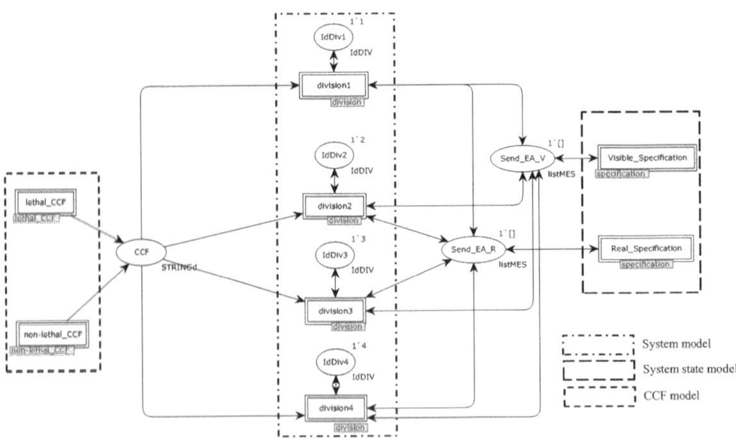

Figure 14.7. *High level colored Petri net of the I&C system*

14.2.2.3.1. CCF generating using Atwood model

Non-lethal CCF are modeled by the CPN sub-net as shown in Figure 14.8. It corresponds to the substitution transition "DCC_non_lethal" of the Figure 14.7. The place "Nb-carte" contains the number of electronic boards N in the system. The firing of transition "Save" maintains the number of electronic boards N of the system in the place "SNb_carte" and sets N tokens in the place "nb_carteu". The transition "proba" is fired N times. The function "defdcc()" draws a random value using a uniform distribution law of probabilities in the interval [0, 1]. If the value drawn is lower than conditional probability p, the considered board will be shock sensitive. The returned value will be 1, otherwise 0.

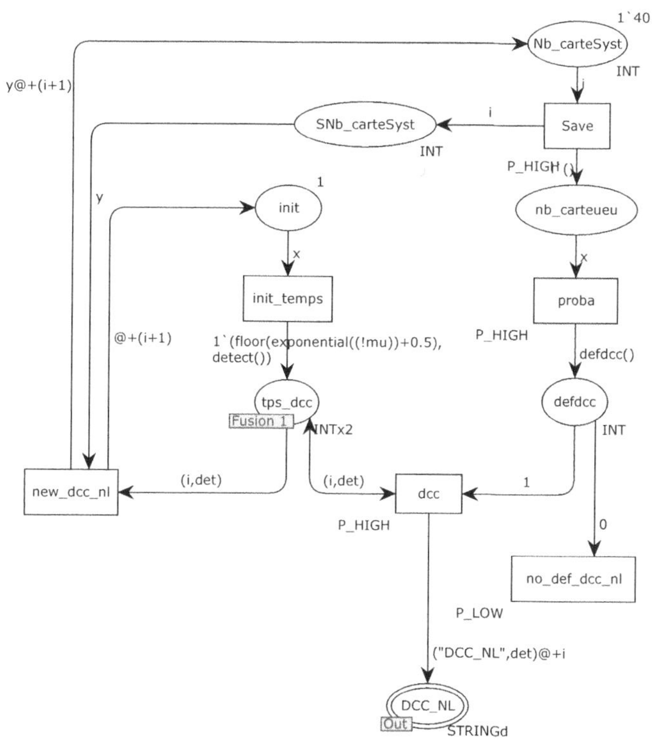

Figure 14.8. *CPN subnet modeling the non-lethal CCF*

The firing of “init_temps” transition allows us to determine the occurrence time of the non-lethal shock using the function “floor (exponential(!mu)+0.5)” and, at the same time, specifying if it is detected or not by a self-test. This is done using the function “detect()”. This function draws a random value using a uniform distribution law of probabilities in the interval [0, 1]. If the value is lower than the coverage rate of self-tests (t_C), the non-lethal shock is detected. In this case, the function returns the value 1, otherwise 0.

The firing of the transition “dcc” assigns the occurrence time of CCF and the variable of failure detection at each token of the system that is shock sensitive.

The transition “no_def_dcc” allows us to remove the tokens representing boards not being shock sensitive.

The transition “new_dcc” allows us to generate the next occurrence time of the non-lethal shock and to redefine the number of boards being shock sensitive.

Lethal CCF are modeled by the CPN subnet as shown in Figure 14.9. It corresponds to the substitution transition “DCC_lethal” in Figure 14.7. The firing of the transition “gene_dcc_l” allows us to determine the occurrence time of the lethal shock using the function “floor (exponential(!omega)+0.5)”. A lethal CCF affects all the components (electronic boards) of the system and is always detected online. Thus, there are N timed tokens issued with a color (’DCC-NL’,1) (1 for detection). The next occurrence time for a new lethal CCF is calculated since the previous is realized.

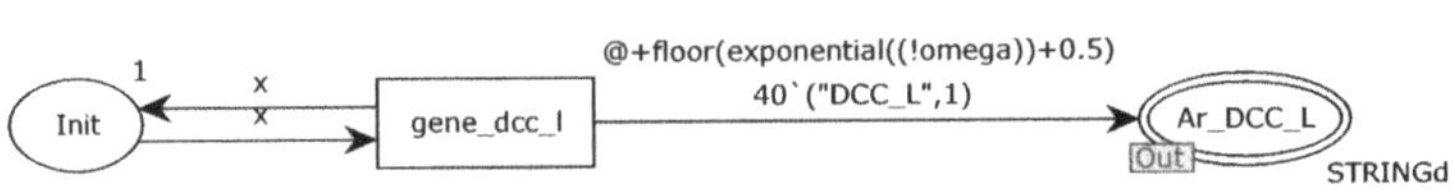

Figure 14.9. *CPN subnet modeling the lethal CCF*

The Atwood model introduces three parameters to estimate: p, μ and ω. Two approaches, analytic or by simulation, can be used to access their estimation. As we focus here on PN modeling, the analytical approach will not be presented (see [BRI 14, DEL 15b, DEL 15a]).

Monte-Carlo simulation approach for estimation of CCF model parameters

We introduce the ratio between the occurrence rate of independent failures λ_{IND} and the total failure rate λ_{TOT} of a component (an electronic board):

$$\gamma = \frac{\lambda_{IND}}{\lambda_{TOT}} = (1-\alpha).\sum_{k=1}^{n} C_{N-1}^{k-1}.p^{k}.(1-p)^{N-k} - \beta_{lethal} \quad [14.7]$$

$$\gamma = 1 - \alpha.p - \beta_{lethal} \quad [14.8]$$

We also introduce E_i, which represents the number of independent failure events realized in 10 years of simulations. This number is obtained by only simulating the independent failures in the PN model and their occurrences are counted. It allows us to obtain from equation [14.9], the average independent failure rate λ_{IND} for a board in the system. The different rates can be estimated from the following equations:

$$\lambda_{IND} = \frac{E_i}{10years \times N} \quad [14.9]$$

$$\mu = \frac{\alpha.\lambda_{IND}}{\gamma} \quad [14.10]$$

$$\omega = \frac{\beta_{lethal}.\lambda_{IND}}{\gamma} \quad [14.11]$$

The expected value of E_i is equal to 8.28 and the value obtained for independent failures rate is: $\lambda_{IND} = 2,36.E^{-6}.h^{-1}$. The results obtained by applying this approach are presented in Table 14.3.

Although an analytical approach may conduct to the same result [BRI 14, DEL 15b], the Monte-Carlo simulation approach is able to revisit some assumptions supporting the binomial failure rate model underlying the Atwood shock model, especially in the case of digital I&C systems including software.

p	γ	λ_{TOT}	μ	ω
0.2	0.995	2.37 10^{-6}	9.62 10^{-7}	1.18 10^{-8}
0.33	0.995	2.37 10^{-6}	9.62 10^{-7}	1.18 10^{-8}
0.5	0.995	2.37 10^{-6}	9.62 10^{-7}	1.18 10^{-8}

Table 14.3. *The values of occurrence frequencies μ and ω for non-lethal and lethal shocks as a function of p*

14.2.2.3.2. Introduction of asymmetry in CCF propagation

For a realistic modeling of CCFs and of their impact on the systems dependability, it is interesting to assume that some non-lethal shocks predominantly affect some k components among all the N components of the system, where $1 \leq k \leq N$ (e.g. only some types of electronic boards belonging to the I&C systems). We propose to modify the generation of non-lethal CCF for simulating the asymmetries in their propagation, in other words considering "oriented" CCF.

In a system with N components, a non-lethal shock affects the boards with a conditional probability p. The expected number of affected components is $N \times p$. The set of N components can be divided into two subsets A (SSA) and B (SSB), containing, respectively, a and b components, such as $a + b = N$. Let x be (respectively y) the probability that a component of SSA (respectively SSB) is affected by a non-lethal shock. From reasoning on mathematical expectation, we have:

$$ax + by = Np$$

To determine x and y we choose how a CCF will affect the SSA (p_1) and SSB (p_2), such as $p_1 + p_2 = 1$. Thus, we obtain:

$$x = \frac{N.p.p_1}{a}$$

$$y = \frac{N.p.p_2}{b}$$

The solutions of these equations, x and y, are the probabilities of non-lethal shock that will be incurred by the components of subsystems SSA and respectively SSB.

14.2.2.3.3. System modeling by CPN

The hierarchical and modular aspects of CPNs are exploited to develop the CPN model of the I&C system. Thus, each of its divisions is modeled by means of a substitution transition at the high level of the system, in Figure 14.7. A division includes five APU, each of which includes the correct number of electronic boards C1 and C2. The CCF are transmitted to the APU and to the boards by means of CPN places (socket places). The state of an APU is determined by the state of its boards (available or unavailable). Once the state of a board changes, the new state of the APU is sent to the specification that determines the state of the whole I&C system.

The CPN model of an electronic board represented in Figure 14.10 has three possible tangibles states: operational ("Marche" place), failed and non-detected ("detection" place) and being repaired ("reparation" place). When the board is operational, it can change its state if one of the following four events occurs:

– independent failure detected online by a self-test (modeled by the Petri sub-net on the right side of the model);

– independent failure detected offline in periodical-tests (modeled by the sub-Petri net on the left side of the model);

– non-lethal CCF detected online and lethal CCF (modeled by the tokens received by "AR_DCC" place from the CCF models of Figures 14.8 and 14.9);

– non-lethal CCF detected offline in periodical-tests (also modeled by the tokens received by "AR_DCC" place from the CCF model of Figure 14.8).

If the failures are detected during the periodic tests, the "rep" transition is fired and the board changes its state to the repair state. If the failures are detected online, the board changes its state immediately from the operational to the repair state. The repair time is calculated using the function "floor (exponential(1/!MTTR)+1.0)". As soon as the board is repaired, its state changes immediately to the operational state. The next occurrence times of the independent failures detected online and offline are also calculated.

This CPN model of electronic boards is generic (Figure 14.10). For all boards, only the numerical values of parameters (failure and repair rates) are different.

14.2.2.3.4. System state modeling by CPN

Using the information about the state (available/unavailable) of the APU i ($i \in [0, 4]$) of the division j ($j \in [1, 4]$), it is possible to determine the state of the I&C system during the Monte-Carlo simulation using the CPN of Figure 14.11. The system state is represented by a token whose color is composed of five Booleans, each of them representing the state of one APU. The different configurations on the transitions' guards allow us to define the conditions of availability/unavailability of the system.

The entire CPN model obtained for the I&C system has 685 transitions and 504 places. Even if the model size is large enough, the use of hierarchy and color concepts resulted in a modular and readable model obtained through the instantiation of generic templates. We can note that an equivalent classical PN model for the same I&C system should have several thousand places and transitions. The obtained CPN model allows the probabilistic dependability assessment of the system.

14.2.3. *Probabilistic dependability assessment*

14.2.3.1. *Probability of failure on demand*

Since the I&C system is a safety related system, we are interested in assesing only the probability of failure on demand (PFD). This is in accordance with the IEC61508 standard [IEC 10] and its declination IEC61513 [IEC 11], which address the functional safety of digital systems. We define two kinds of PFD:

– the "real PFD" that represents the I&C system unavailability starting at the time instant where the electronic boards are out of their operational state;

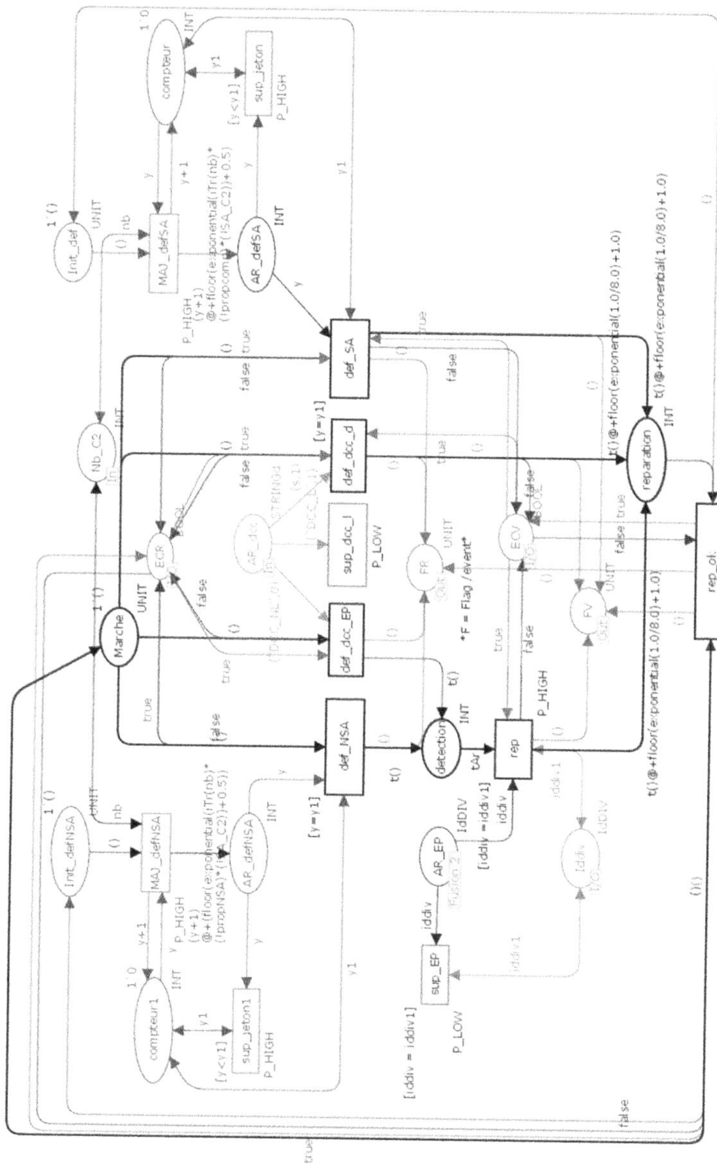

Figure 14.10. *CPN sub-net of an electronic board*

– the "visible PFD" that represents the system unavailability only starting with the time instant where the boards failure is detected. The time between the occurrence of non-detected failures and their detection by a periodic-test is not considered.

The definition of these two kinds of PFD gives an access to the difference between the real state of the system and the state observed by plant operators during the system operation.

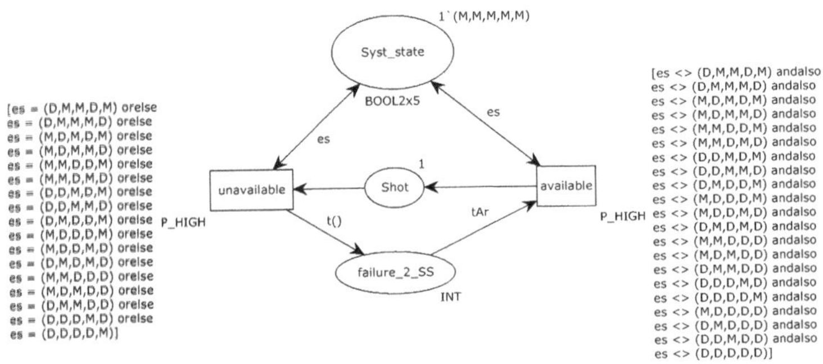

Figure 14.11. *CPN sub-net to determine the state of the whole I&C system (available or unavailable)*

The PFD is estimated using Monte-Carlo simulation by the following equation:

$$PFD = \frac{System\_unavailability\_time}{Mission\_time}$$

The results presented in Table 14.4 were obtained after 10,000 histories over 10 years. The 95% confidence interval for all the results presented in Table CC3 is also estimated and its value is $< 10^{-6}$.

According to [IEC 11], "there is no equivalent scheme to the reliability/risk reduction safety integrity levels (SIL) proposed in IEC 61508 in common use in the nuclear sector". Consequently, we do not associate a SIL level for the system under study.

Indicators	p	Average
real PFD	0.2	$2.5\ 10^{-5}$
visible PFD	0.2	$<\ 10^{-6}$
real PFD	0.33	$2.73\ 10^{-4}$
visible PFD	0.33	$1.0\ 10^{-6}$
real PFD	0.5	$5.4\ 10^{-4}$
visible PFD	0.5	$4.0\ 10^{-6}$

Table 14.4. *Real and visible PFD of the protection I&C system for different values of* p

If the conditional probability of failure of one component in a non-lethal shock p increases, there will be more components affected by a shock and the PFD also increases. The difference between the real PFD and the visible PFD is about 10 to 100.

14.2.3.2. *Influence of the oriented non-lethal CCF*

To assess the impact of asymmetric propagation of non-lethal CCF, we consider the orientation of CCF on the two subsystems SSA and SSB, using the approach presented in section 14.2.2.3.1.

The conditional probability of failure of one component in a non-lethal shock is $p = 0.2$. The frequency of non-lethal shocks is arbitrarily fixed at one shock per year, i.e. $\mu = 1.14 \times 10^{-1} \times h^{-1}$. The lethal shocks are not represented. Ten thousand trajectories over 10 years are also realized and the occurrence of system failure ends the trajectory simulation. Table 14.5 shows the results. The sum of the failures' combinations (columns) is equal to the total number of histories (10,000) for a fixed CCF orientation.

We can observe that the more a subsystem is preferred (according to p_1 and p_2), the more the MTTFF increases. The logic vote explains this phenomenon. We also observe that independent failure combinations without CCF rarely lead to system failure, due to the high level of redundancy. Independent failures detected offline lead more easily to system failure than those detected online. This aspect can be improved by increasing the frequency of periodic tests or the coverage rate of self-tests.

CCF orientation p_1 SSA	0.1	0.2	0.3
p_2 SSB	0.9	0.8	0.7
only CCF	9,404	9,723	9,787
indep. SA failures	3	0	1
indep. NA failures	21	9	9
indep. SA and NSA	77	46	42
CCF & indep. SA failures	8	6	1
CCF & indep. NSA failures	487	216	160
MTTFF (in h)	49,586	28,921	22,056

Table 14.5. *Combinations of failed boards leading to the system downtime depending on p_1 and p_2*

14.2.4. *Conclusion*

The CPNs prove relevant for the representation of digital I&C systems and the assessment of their dependability indicators considering CCF. The shock Atwood model has been transposed in PN. It allows us to take into account not only the independent failures and the lethal CCF, but also the non-lethal CCF. An extension of this model was proposed to represent the asymmetric propagation of non-lethal CCF on privileged axes. This allows us to relax the assumption of the random repartition of non-lethal shocks on all components and to represent the effects of diversity.

The extremely low volume of operating experience related to the protection I&C systems makes it difficult to estimate the α factors representing the conditional probability of failures on demand of the k components in a group of N components when a CCF occurs. This CPN model could be used to simulate operating experience for obtaining more data, which make a more accurate estimation of these factors. Another interesting axis concerns the estimation of the three parameters of the Atwood model.

15

Impact of Failures on System Performances

15.1. Reliability evaluation of networked control systems

15.1.1. *Statement of the problem*

Technology advances and the availability of network connectivity have prompted the idea of introducing network facilities to control systems. Such systems are called NCSs: their sensors, actuators, control units, diagnosis and supervision units are connected through communication networks as shown in Figure 15.1.

In a classical (network-free) system, the controller periodically performs the synchronization of the functions: input data acquisition via sensors, computing of the next control order as a function of this data and the previous internal state and finally control order transfer to actuators. Within an NCS, sensors themselves periodically sample the data input and send it to the controller via the network. After the computing of the control order, the controller sends it to the actuator *via* the network. The introduction of the network in the control loop is then the origin of multiple problems as the control of synchronization becomes very dependent on this network. Delayed sensing and actuation, time-varying transmission periods, network schedulability, network-induced delay, data transmission errors, data packet loss and so on, are of concern [SOG 06].

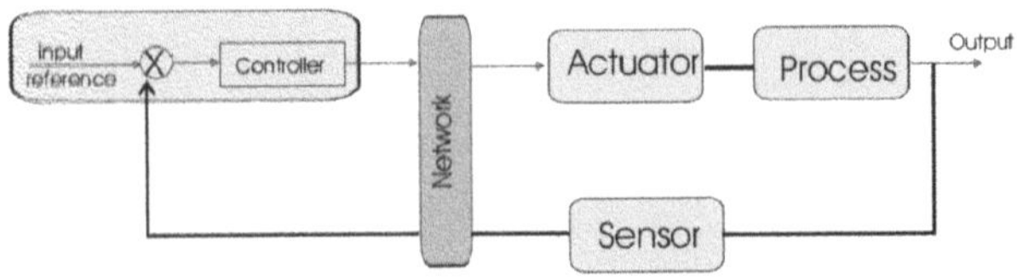

Figure 15.1. *Structure of an NCS*

Performance predictive analysis of NCS becomes very difficult due to the multiplicity of disturbance sources and sometimes due to the random behavior of the network protocol. Resorting to a simulation approach then becomes a reasonable means of study. In [ZHA 01], Zhang *et al.* used the Case Western Reserve University campus-wide network (CWRUnet) to simulate the NCS environment and analyze the effect of network-induced delay and packet loss. Branicky *et al.* [BRA 03] developed a simulation tool that combines dynamic-system simulation for the control agents and environment with packet-level network simulation for the communications by extending network system-2 (NS2).

With regard to reliability and dependability, the problem arises in an identical way. In addition to catalectic failures such as the breaking of the network medium whose effect is identical to a component failure and consequently will be encountered in classical reliability studies, it is important to consider the problems previously described. In network-oriented approaches, faults affecting the network may have various effects on the control system, depending on the state of the system and on the failure occurrence. For example, in a closed loop, loss of messages in the network during the transient state does not have the same effect as in the steady state. Another important problem is the transient character of some faults that can even be fugitive. The consequences are that in the general case an analytical study of the dependability is not easily achievable. Otherwise, remembering that control loop devices have a disturbance rejection capacity and even the ability to compensate some consequences of component failures, it becomes necessary to reconsider the concept of failures of the whole system.

It appears therefore that regarding the multiplicity and complexity of the relevant problems, reliability predictive assessment of NCS is more a matter of using approaches based on simulation than of analytical formal study. In his Ph. D. thesis [GHO 08a, GHO 08b, GHO 11], R. Ghostine[1] presents a Monte-Carlo approach for the statistical evaluation of the reliability parameters due to many simulations (or stories), and makes it possible to obtain results. This work is presented in the following sections.

15.1.2. *Reliability criteria of an NCS*

The IEC 60050-191 standard [IEC 90] defines the reliability as "the ability of an item to perform a required function under given conditions for a given time interval" and its measure "the probability that an item can perform a required function under given conditions for a given time interval". In traditional approaches, the mission or function of the entity is considered as fully performed or failing so that Boolean models are convenient to define the reliability function (or structure function). In case of feedback controlled systems, we can define the mission of the system as being the control of a physical variable with a given quality. The quality of control is defined by some criteria as for example, for the response to an input step, a limited overshoot, minimal rising and response times, and no trend to instability. Any behavior of the system which does not respect the specifications on these criteria must be regarded as a failure of the system. This may be justified by the fact that in many industrial systems, an excessive value of a physical variable may be dangerous for a part of the system and then impact the reliability of the whole system (for example over current in an electric drive, over torque on a shaft, over temperature in a semiconductor, etc.). Ghostine *et al.* [GHO 08a, GHO 11] introduced the "failure by a criterion" as being the probability that this criterion is

1 Ph.D. thesis performed under the supervision of Prs. J-F. Aubry and J-M. Thiriet.

not respected. The system will be considered as failed when at least one criterion (overshoot, response time and rising time) fails.

As an example, let us define two types of failure by criterion, the failure by overshoot and the failure by stability.

– Failure by overshoot: this means that the overshoot in the response to a step input is too high. A maximum gap D_{ov} between the output and the ideal response to a step must be defined. In case of transient faults, the system may affect its response to a step input if at a given time the overshoot is greater than the set-point D_{ov}. So we can consider that a failure by overshoot occurred in the system. We can physically justify this choice since in many industrial applications the overshoot of certain variables, such as for example an overvoltage in a semiconductor, can be dangerous or even fatal.

– Failure by stability: the purpose here is to identify situations where the disturbances have a trend to make the system unstable [BAB 03]. To define the failure by stability, supervision of the successive peaks of the response to a step input in the presence of transient faults is performed. If a trend to oscillation is observed, we can conclude system unstability. More precisely, the criterion may be as follows: if three successive amplitudes of peaks (in absolute value) are increasing, the system is declared in failure by stability.

15.1.3. *Elements of modeling*

15.1.3.1. *Hierarchical general model*

The PN of the entire NCS is presented in Figure 15.2. All transitions are substitution transitions (double line bordered transitions) with which a sub-PN is associated in order to describe the detailed behavior of each component. The corresponding sub-PN of substitution transitions are indicated by rectangular tags positioned next to them, and for reasons of simplicity these sub-PN have the same name as the transitions. The NCS is composed of:

– a physical process that must be controlled;

– a sensor that measures the output variable of a process;

– a controller that develops a new control value according to the measured value obtained from the sensor and to the set-point value representing the desired output value for the process;

– an actuator that acts on the process according to the control value received from the controller;

– a network that transmits messages from sensor to controller and from controller to actuator.

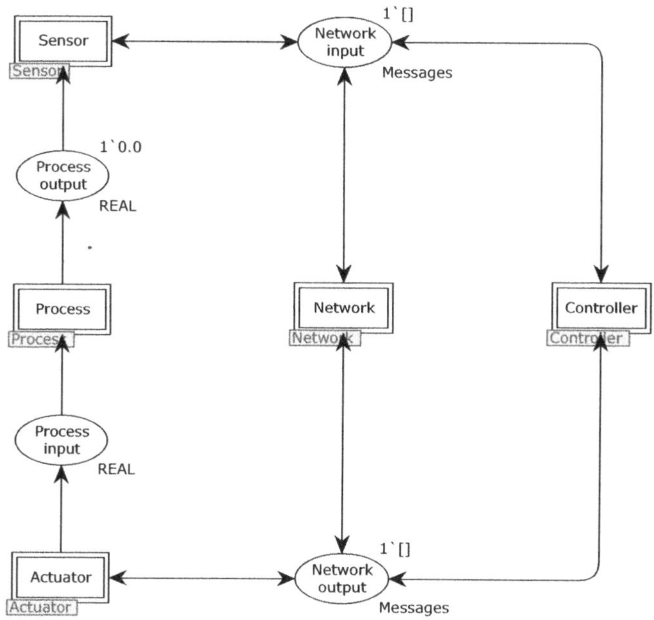

Figure 15.2. *System-level CPN model*

15.1.3.2. *Process model*

The CPN model representing the physical process is presented in Figure 15.3. According to its actual state variable x_k (given by the value of token in place "Process state") and control value received from the actuator u_k (given by the value of token in place "Process input"), it develops the next value of its state x_{k+1} and its output value y_k. The next value of its state x_{k+1} is stored by the token added to the place "Process state" and the output value y_k is sent to the "Process output" place. The places "Process input" and "Process state" are respectively input and output port places (represented by rectangular port tags positioned next to them) that are linked to the corresponding places connected to the substitution transition "Process" in the high level of CPN (Figure 15.2).

The previously described behavior of the physical process is modeled by the firing of transition "Process". When this transition is fired, the function $processState(x_k, u_k)$ presented in the code segment of the transition is executed. This function implements a specific model of the physical process:

$$\begin{cases} x_{k+1} & = & f(x_k, u_k) \\ y_k & = & g(x_k, u_k), \end{cases}$$

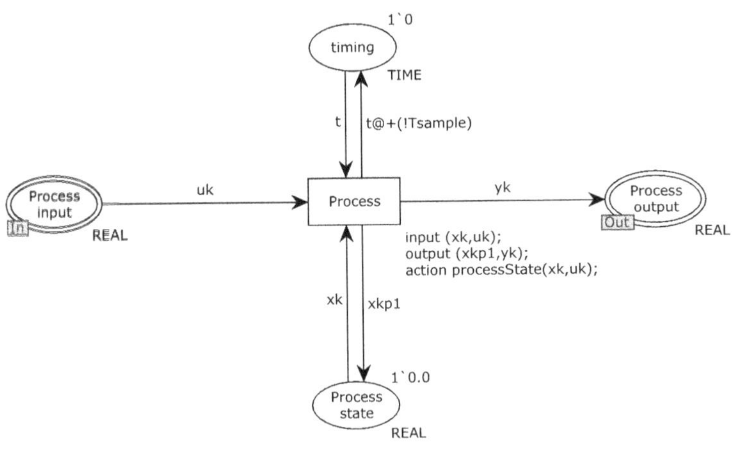

Figure 15.3. *Process CPN model*

The CPN model of Figure 15.3 is generic and can be used unchanged to describe any controlled physical process. Only the expressions of $f(\cdot)$ and $g(\cdot)$ in the function $processState(x_k, u_k)$ must be changed to describe any physical process.

The transition "Process" is fired periodically with a sample period given by the variable $Tsample$. This is assured by the timed token of the place "timing". At each firing of transition "Process", the timestamp of this token (describing the instant time when it will be available for enabling the next transition firing) is changed with the value of the sample period. The sample period is modeled by a global reference variable of CPN $Tsample$. The inscription $!Tsample$ gives access to the value of the reference variable $Tsample$.

15.1.3.3. *Sensor model*

The CPN model representing the sensor component of NCS is presented in Figure 15.4. When the transition "Sensor" fires the output value of the physical process (given by the token value of the place "Process output" linked to the corresponding place of the physical process) it is measured and a new message is generated for transmitting this measured value to the controller using the network.

A message is a triple $(id\_mess, data, error\_code)$ containing one identifier, the data that must be transmitted and an error code that will be used to determine if errors occur during the network transmission. The message is generated by the function $new\_message(!id\_S, yk)$. This function has two parameters and, first, it generates the error code according to a given error code detection mechanism (such as cyclic redundancy check, CRC code), and second, it generates a message represented by the corresponding triple. Afterward, the message is added to the messages queue that must be transmitted by the network. For this, the current queue is removed for the place "network input" and replaced by an updated queue that adds, at the end of the current queue, the new generated message (inscription $queue\hat{}\hat{}[new\_message(!id\_S, yk)]$). This is realized only if the sensor is in its operating state. The sensor state is described by the value of token stored in the place "Sensor state". This sensor state is determined by a series of self-tests that are carried out with a period given by the global reference variable $TselfTest$. This is modeled by the place "timing" and the transition "Sensor failure" that fires periodically according to the timestamp of the token situated in the place "timing". This mechanism is similar to the mechanism that models the sampling of physical process. When the transition "Sensor failure" fires, the function $failure(!s)$ is executed. This is a generic function that implements self-tests of the sensor s and according to the results of these self-tests, the state of the sensor is updated. The function $failure(!s)$ can also model the occurrence failure of the sensor according to a specific probability law describing its failure. In this last case, the failure occurrence is tested periodically by the firing of the "Sensor failure" transition in order to verify if it is already in its operating state.

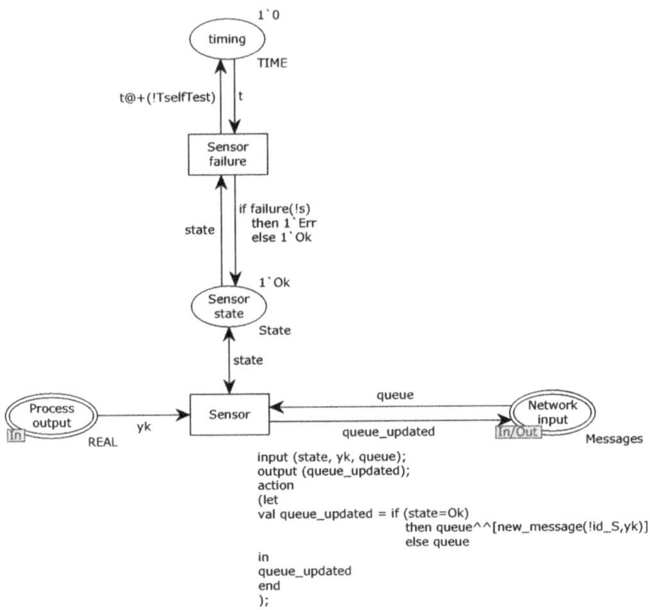

Figure 15.4. *Sensor CPN model*

The CPN model of the sensor presented in Figure 15.4 is a generic one. It can be used unchanged to model a given sensor, both in terms of failure occurrence and of network message generation. As in the controller area network (CAN) protocol, the message has a unique identifier, but any other network protocol can be modeled by changing only the function $new\_message(\cdot)$.

15.1.3.4. *Controller model*

The CPN model representing the controller of NCS is presented in Figure 15.5. If the controller is in its operating state ($state = Ok$) and if it is concerned with a message received by the network, the transition "controller" fires in order to execute the controller tasks. Due to unique identifier of a message, the network stations (controller and actuator), that are connected to the transmission medium of network, recognize and process messages that concern them. This is modeled by the second part of the guard of the transition "controller" ($\#1$ $message) = !id\_S$ representing the test that the identifier (the first element of a triple that models a message $\#1$ $message$) is equal to the identifier of the message generated by the sensor $!id\_S$ to the controller. When the transition "controller" fires, the received message is removed from the "network output" place. The data contained by the message received from the sensor (given by the second-term of the message $\#2$ $message$) and the desired value from the physical process output (given by the value of reference variable $!setpoint$ is used to generate the control value according to the corresponding algorithm of control such as well-known proportional-integral-derivative (PID) or other digital controllers. This is realized by the execution of the function $generate\_control$ $(!setpoint,$ $\#2$ $message)$. This function determines the new value of the control variable u_k. The obtained value of u_k is used to generate a new network message for the actuator $new\_message(!id\_A, uk)$ and afterward, this message is added to the queue of messages that must be transmitted by the network $output\_queue$^^$[new\_message(!id\_A, uk)]$. The state of the controller is modeled and updated by a similar PN, such as for the sensor.

15.1.3.5. *Actuator model*

The CPN model representing the actuator component of NCS is presented in Figure 15.6 and is similar to the sensor and controller models. The actuator state is modeled and updated similar to previous models. If the actuator is in its operating state ($state = Ok$) and if it is concerned by a message received by the network ($\#1$ $message) =$ $!id\_A$, the transition "actuator" fires in order to execute the actuator tasks. The received message is removed from the "network output" place and the data representing the new value of the control variable, determined and sent by the controller (given by the second-term of the message $\#2\ message$), is applied to the physical process (a token with the corresponding value is added to the place "process input" that is linked to the corresponding place of the process CPN model of Figure 15.3).

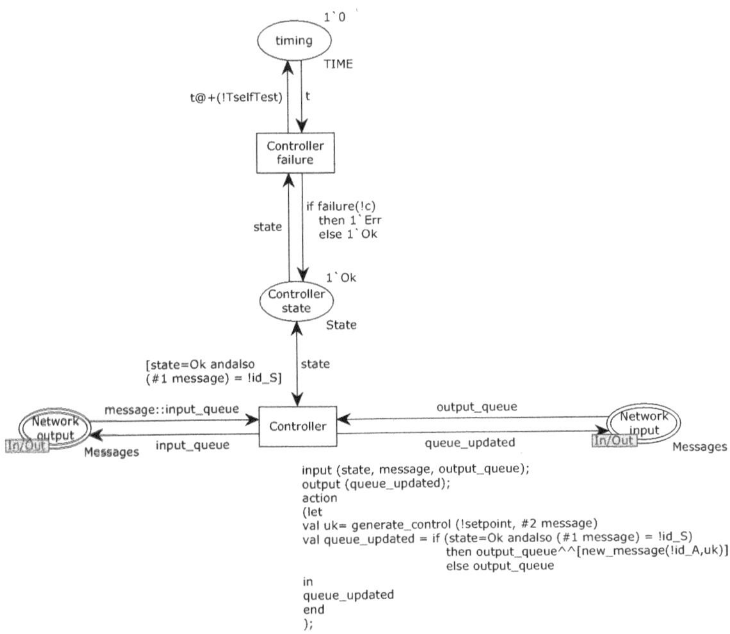

Figure 15.5. *Controller CPN model*

15.1.3.6. *Network model*

The CPN model representing the communication network of NCS is presented in Figure 15.7. If the network is available and if there are messages to be sent, the transition "begin transmission" fires. The condition to start a transmission (the queue of waiting messages is not empty) is modeled by the guard of this transition $queue\ != [\ ]$. When this transition fires, first, the highest priority message that must be transmitted is determined by the function $get\_priority\_message(queue)$. Second, the state of the queue with waiting messages for transmission is updated, by removing the current message that is now being transmitted (this is realized by the function $update\_queue(queue, message)$. This updated queue is refiled in the place "network input" so to that other system components can realize their tasks in parallel to the current transmission. The message is added to the place "message transmission" from which the "error detection" transition fires. The detection process of an error is modeled by the function $error\_detection(message)$. This function can detect an error based on the third part of a message representing the error code. This function is a generic function, it implements a specific error code detection mechanism as needed (e.g. parity, CRC, etc.), without modification of the CPN model of Figure 15.7. According to the result obtained (variable $error$) of this error detection mechanism, the message is added to the place "not erroneous message" (if no error is detected) or to the place "error" (if an error is detected).

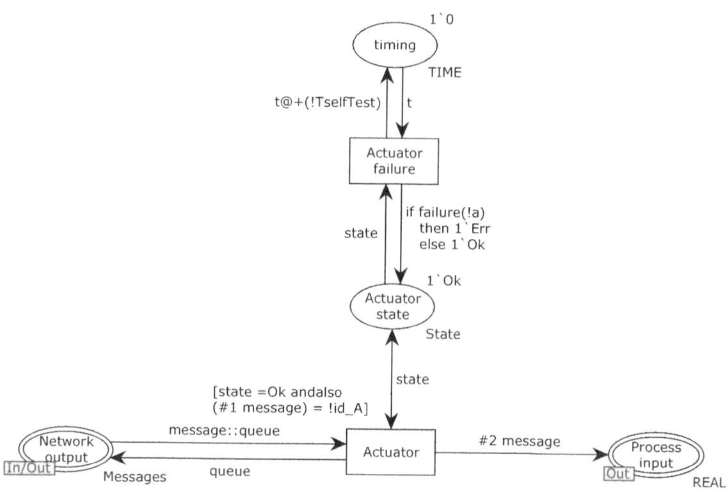

Figure 15.6. *Actuator CPN model*

The token added to the place "not erroneous message" is a timed token (its timestamp is determined by the function $transmission\_delay()$). The timestamp of this token represents the necessary delay to finish the transmission. This function *transmission_delay*() determines the transmission delay according to the hardware configuration of the network and used protocol. At the end of the transmission, the message is added to the queue of the sent messages at the network output. The network is available again for a new transmission (token added in the place "network available").

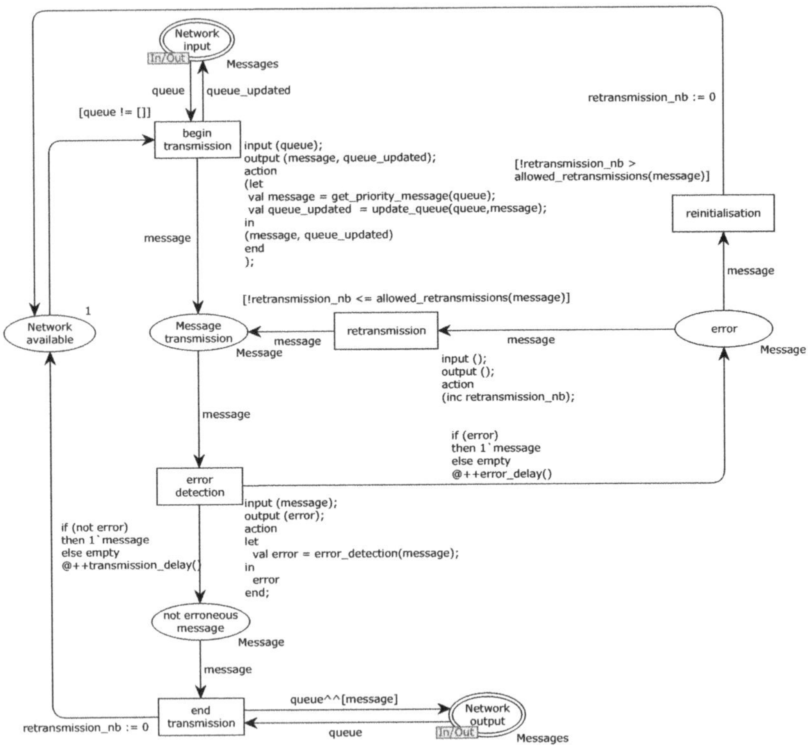

Figure 15.7. *Network CPN model*

The token added to the place "error" is a timed token (its timestamp is determined by the function *error_delay*()). The timestamp of this token represents the necessary delay to restart the transmission. This function *error_delay*() determines this delay according to the protocol used. At the end of this delay, the transition "retransmission" occurs, if the number of retransmissions does not exceed a maximal allowed number of retransmissions (modeled by transition guard

$!retransmission\_nb <= allowed\_retransmissions(message))$. The number of allowed retransmissions is specific to each message and it is determined by the function $allowed\_retransmissions(message)$. According to the communication protocol used in the network, this allowed number of retransmissions can be fixed in advance or determined as a ratio between the maximum duration that is allocated to achieve transmission and the message size. If the allowed retransmission number is exceeded, then the message is lost and the transmission is reinitialized (a token is added to place "network available") in order to send a new message. The variable $retransmission\_nb$ representing the number of attempted transmissions is also reset.

15.1.4. *Simulation and results*

In order to obtain significant evaluations, it is necessary to produce several simulations to obtain a degree of confidence of the estimated variables. Let us recall that with the Monte-Carlo approach, it is not guaranteed to have the exact values of the parameters to be evaluated. In fact, the method guarantees only the probability that the number of simulated stories N ensures a result with a given precision. To define the confidence in the results, two parameters are used: a precision interval and a percentage of values which belongs to this interval. Each time a new history i (i being the number of the history) is launched, a new average value V_i is determined and the difference $d_i = V_i - V_{i-1}$ is calculated. If d_i belongs to the precision interval, the number of histories belonging to the precision interval increases. If the percentage of histories belonging to the precision interval reaches a given threshold, the simulation stops. For example, an accuracy of 5% of the precision interval and 95% of the percentage of membership requires that 95% of the calculated d_i be lower than 75% of the last average. In the next sections the cases of the failure by overshoot and the failure by stability are studied. Depending on the system, other criteria can of course be studied.

The results presented here were obtained by simulation of models in the Möbius software tool based on stochastic activity networks, a formalism related to PNs. R. Ghostine presented in his Ph.D. thesis report [GHO 08a] the equivalence between the Möbius and CPN tools models of a NCS.

15.1.4.1. *Influence of a variable delay on the failure by overshoot*

The response to a step input (one history) is simulated by simultaneously introducing a random delay on the delivery of information between sensor and controller on the one hand and controller and actuator on the other hand. A series of such simulations is performed and for each of them, the fulfillment of the overshoot criterion is verified or not. With each new history, the average number of times where the criterion is not satisfied (overshoot higher than the threshold) is calculated. The simulation is stopped when the difference between two average successive values is lower than 75% of the last average, and this at least for 90% of the stories.

Figure 15.8 illustrates the relationship between the variable delay which expresses the quality of service of the network and the failure by overshoot, representative of the quality of control. The curves show the change of the probability of failure by overshoot according to the D_{ov} variation. The three curves correspond to three classes of variable delay. The delay is uniformly distributed between 10% and 30% of the sampling period for the first case, between 15% and 35% for the second case and between 20% and 40% for the third case. These values are taken to illustrate the influence of the variable delay. They are however realistic; for the first case, if the communications protocol manages the access in a priority way, we can for example imagine a situation where a higher priority message (which will be sent before the messages relating to the control loop) introduces a delay of 10% of the sampling period (125 bits for a rate of 125,000 bit/s); after this delay, the network will be temporarily unreachable for one random time between 0% and 20%. Other causes can generate this type of delay: for example in non-deterministic open networks where the traffic can vary in a random way.

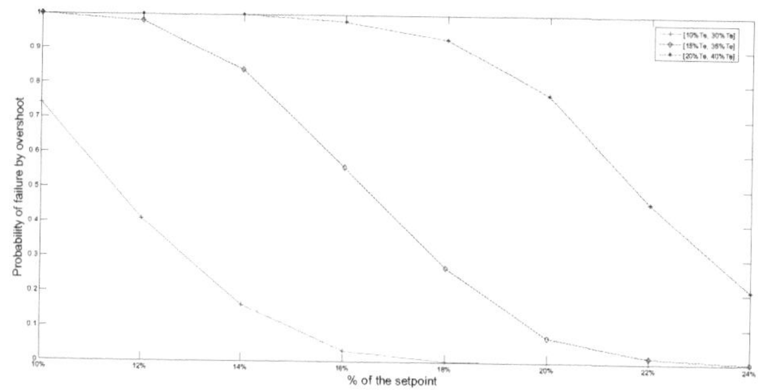

Figure 15.8. *Probability of failure by overshoot in the presence of variable delays, the x-axis represents the constraint of the D_{ov} threshold (expressed in % of the setpoint), the y-axis represents the value of the probability of failure*

15.1.4.2. *Influence of message loss on the failure by overshoot*

The same step input is applied to estimate the influence of the loss of messages. The simulation of the response to a step input (history) is repeated by simultaneously introducing random losses on the information exchange between sensor and controller on the one hand, and controller and actuator on the other hand. Monte-Carlo simulation is used for the evaluation of the probability of failure by overshoot and the same criterion to stop is used as in the case of the variable delays (5% of precision and 90% of percentage of membership). Figure 15.9 shows the probability of failure as a function of the threshold constraint for various probabilities of loss.

These results show the existence of a relationship between the probability of the losses of a message and the failure by overshoot of the system. As we pointed out concerning the variable delay, we can wonder about the origin of the losses to characterize their probability of occurrence according to the environment and the traffic circulating on the network. We can for example take into account the probability occurrence of two lost consecutive messages (this possibility is plausible in the case of wireless networks).

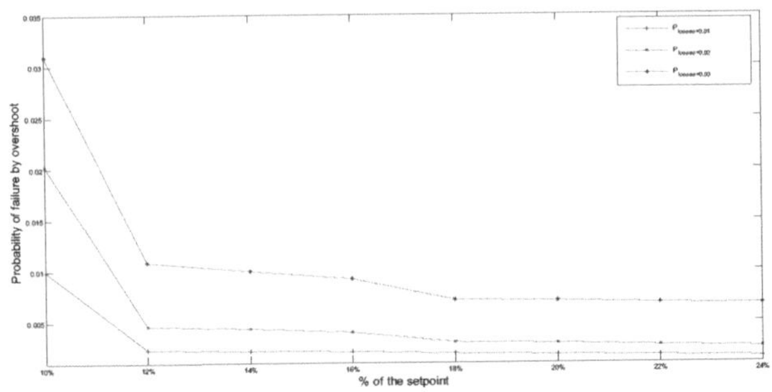

Figure 15.9. *Probability of failure by overshoot in the presence of losses of information, the x-axis represents the constraint of the D_{ov} threshold (expressed in % of the setpoint), the y-axis represents the value of the probability of failure*

15.1.4.3. *Influence of the superposition of variable delays and loss of messages on the failure by overshoot*

The two previous sections considered only one source of fault and made it possible to highlight the relationship between the source of fault and the failure by overshoot. The following part aims to highlight the simultaneous influence of these two sources on the failure by overshoot. The delay is uniformly chosen between 10% and 30% of the sampling period and the probability of loss is fixed at 0.01. In Figure 15.10, the three curves correspond to the three cases: (1) variable delay, (2) losses of messages and (3) superposition of delay and losses of messages.

A probability of message loss of 0.01 on a network is a rather high probability, in practice this probability is smaller; while a delay varying uniformly between 10% and 30% is more realistic. With these values we find that the overshoot is much more sensitive to the variation of the delay than to the loss of the messages (the probability for the setpoint of being overtaken by 10% is about 0.7 for the variable delay hypothesis and of 0.009 for the loss hypothesis with a probability of message 0.01). The results of simulation show that there are cases where the system is reliable with respect to a single source of failure but not for the superposition of two different sources.

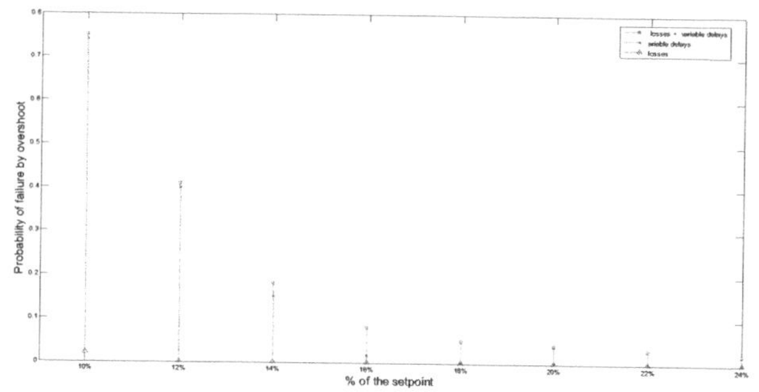

Figure 15.10. *Probability of failure by overshoot in the presence of the losses and the variable delays, the x-axis represents the constraint of the D_{ov} threshold (expressed in % of the setpoint), the y-axis represents the value of the probability of failure*

15.1.4.4. *Failure by stability*

Figure 15.11 shows an example of simulation of the system response to a step input where a trend to instability is observed.

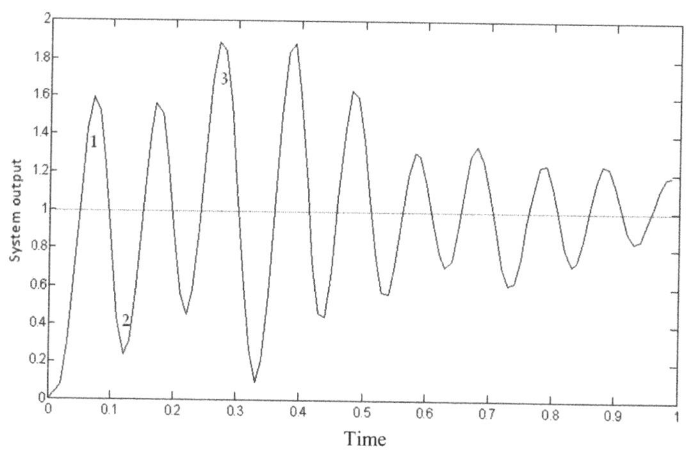

Figure 15.11. *Trend to instability*

Such simulations were performed with the same process as for the previous sections. The probability of failure by stability was performed by the Monte-Carlo method. Table 15.1 shows the results obtained for the probability of failure by stability.

Variable delay ranging $[0, 2.T_s - 0, 6.T_s]$	Probability of failure by stability
Case 1: delay without loss	$0, 0088$
Case 2: delay + prob. of loss = $0, 01$	$0, 0496$
Case 3: delay + prob. of loss = $0, 01$	$0, 1251$
Case 4: delay + prob. of loss = $0, 01$	$0, 228$

Table 15.1. *Probability of failure by stability*

The results are given for a delay varying uniformly between 20% and 60% of the sampling period. In the general case, a first study should be achieved before deciding on the unacceptable number of peaks, this parameter depending on the context of the application. In the present study, a succession of three peaks with increasing amplitude was considered.

15.1.5. *Evaluation of reliability*

According to the definition of reliability given previously, the system fails if one criterion is not satisfied. We only presented two criteria but it is easy to imagine that other criteria may be retained to define the acceptability of the function and then its reliability. Nevertheless, it is obvious that a single fault attributable to the network can lead simultaneously to an overshoot and a trend to unstability for example. It is then not possible to simply add the probabilities measured in separate simulation campaigns. It is evident that a realistic approach would need a simulation with a random injection of all types of listed faults and a simultaneous test of all defined criteria. Such a simulation is possible with the used tools but would lead to a very long execution time.

15.1.6. *Conclusion*

In this example, we saw examples of the influence of failures in the network on the whole control loop performances of a system and the relationship that could link these failures to the notion of reliability of a system. Other studies have been conducted on other factors being possibly influential on the system reaction to failures such as for example the load of the network [GHO 08a]. These possible temporary failures of the system cannot be disconnected from the dependability concept as they can be the origin of propagations leading to the definitive failure of the system or to an unsafe behavior or an accident.

Although the relative complexity of the presented approach (modeling, simulation, etc.) nevertheless has a great interest in the design phase of a system in order for example to help in the improvement of the ability of the controller to reject the perturbations caused by the network or of the capacity of the protocol to tolerate failures.

15.2. Railway signaling

15.2.1. *Introduction*

The free movement of trains throughout European space is hampered among other problems by the diversity of standards relative to the signaling. That is why, with the desire to promote interoperability and consequently further development of the economy in Europe, the European Community [ERT 99] defined a new European railway signaling the standardization of signaling on European cross-border railways. This signaling system called the "European train control system (ETCS)" will have implementation in the more or less short term in the European member states. Several implementation levels of the ETCS has been defined. What we present here concerns the first level [RFF 07].

In order not to disturb the existing railway traffic, the French operator SNCF chooses a superposition of this new signaling to the existing one, the "*Block Automatique Lumineux* (BAL)" (automatic lighting block system). Therefore, the railways receiving the ETCS signaling will allow the circulation of both train types or of interoperable trains (provided with the two signaling recognition systems). However, before the commissioning of such trains it has been important to study the impact in terms of performances (flow, respect of scheduling, robustness to disturbances, etc.) of the coexistence of these two signaling systems ETCS level 1 and BAL. For this purpose, we present the model of a part of a railway track and its two signaling systems. As the complexity of such a system quickly becomes very huge, the modeling tool must be provided with the ability to consider different abstraction levels, at first a macroscopic model of the system, and second an enrichment of this model by taking account of subsystems and components with their detailed requirements. Moreover, time being a critical parameter in railways, the model must be able to account for time evolution anywhere. Colored and timed PNs such as defined in CPN tools [JEN 09] are suitable as still shown in the previous examples. Their ability to introduce random events as we still demonstrated will allow the injection of incidents or failures in the model to test its robustness.

Several publications deal with the modeling of the ETCS system with CPNs. Jansen, Meyer zu Hörste and Schnieder [JAN 98] presented a global approach for the modeling of ETCS functional requirements. A hierarchical CPN model is developed containing the components, the different scenarios of a component's use and, at the lowest hierarchical level, the different implemented functions for each scenario. This model is used in [HOR 98] for a formal validation of the completeness of ETCS requirements. This is achieved by the analysis of the state space (reachability graph). It showed the relevance of CPNs for such studies but stumbles on the combinatorial expansion of the state number. That is why, only a specific scenario of an isolated component is presented.

Lahlou, El-Koursi, Bon and Yim [LAH 06] present a CPN to model the technical interoperability requirements and the operating rules of ETCS. This work can be considered as an extension of the former work. The simulation of the model allows the verification of the respect of interoperability requirements at the level of operating rules.

Barger, Schön and Bouali [BAR 09] use CPNs to model the ERTMS[2]/ETCS requirements. They propose a generation of the reachability graph that will then allow the verification of the safety requirements coherency.

2 The European Railway Traffic Management System.

PNs were also used for the study of the communication between components of the ETCS system. Zimmerman and Hommel [ZIM 03] were interested in the probabilistic dependability assessment of the GSM-R[3] network. This needs the consideration of failures occurrence time in the communication network. An SPN model has been developed to access a probabilistic assessment.

In the following, we present the modeling of a railway track provided with both signaling systems, ETCS and the French lateral signaling BAL as proposed by Lalouette *et al.* [LAL 10b, LAL 10a].

15.2.2. *Interest*

The European Railway Traffic Management System (ERTMS) is the European standard for the control and the signaling of the new railway tracks and the replacement of existing ones. For the deployment beginning of ERTMS in Europe, six corridors has been projected, two of them crossing France needing the equipment of 2,000 km of railway tracks, i.e. more than 4,000 signals superposed to the existing signaling. That is why, a study for the impact evaluation of this new system on the operation of the existing railway network has been engaged.

ERTMS is composed of a control system (ETCS) and a dedicated radio-frequency communication system issued by the GSM standard (GSM-R) both based on the concept of fixed "blocks". These requirements engendered the two following declinations :

– ETCS (level) 1: only composed of ETCS control systems;

– ETCS (level) 2: composed of ETCS control and GSM-R systems.

The East European high speed track is the pilot site for ETCS 2 and the classical tracks will be provided with the ETCS 1 system. This is the case of the track part whose model is presented here.

3 Global System for Mobile communications – Railways.

15.2.3. *Signaling system specifications*

15.2.3.1. *French signaling: BAL*

The BAL signaling type is essentially based on the sequencing of three separate light signals:

– green/free way: the train suffers no restriction;

– yellow/warning: the train will stop at the end of the block;

– red/semaphore or square: the next block is occupied. If the signal state is "semaphore", the train is authorized to enter the block in a special mode called "running on sight", If the signal state is "square", the train must wait for the reopening of the block.

The mode "running on sight" is useful to increase the availability of the line by allowing a train to enter an occupied block under some conditions. The main rule restricts the driver to a 30 km/h speed limit in order to be able to stop as soon as an obstacle is observed on the line. It is a mode for network blocking avoidance.

These light signals serve both for circulation safety by the control of the space between trains, and for itinerary safety by the respect of the switches interlocking.

15.2.3.2. *The European signaling: ETCS level 1*

The ETCS is a cab signaling performing only the spacing function, that is why it rests on the lateral signaling without replacing this one. Instead of giving information on the block's state by a light signal principle, this system imposes on the driver a speed curve to follow. This information is carried on an interface called the "driver machine interface" (DMI) and indicates in real time the maximum authorized speed.

If the train does not respect the maximum speed, it will be taken under control of the ETCS system which will lower the speed to under the limit. It is the principle of movement authority (MA). To compute this MA and speed curve, the train needs information from the track about the state of the signaling and the description of the track used.

Each time this information is received, a new MA and a new speed curve are computed again. These data are received in a punctual way when passing through information points on different locations on the track: at the foot of signals (PI) and upstream them (advanced PI). Each MA has a zero target speed. Figure 15.12 shows an example of MA assignment and speed control curve definition as a function of the BAL signaling.

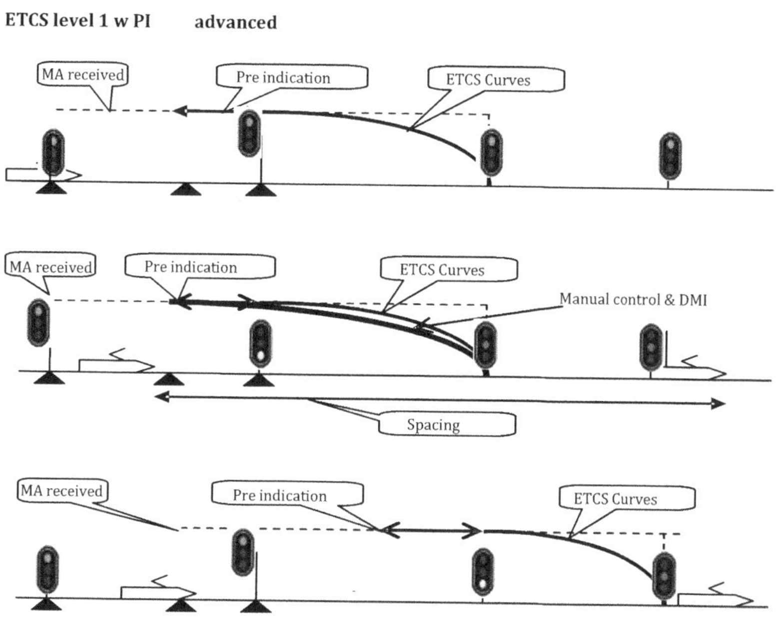

Figure 15.12. *MA assignment as a function of the lateral signaling*

15.2.4. *Elements to be modeled*

The three main elements to be considered are the signaling, the train behavior and the driver behavior.

The two signaling systems being based on different principles, each of them will be modeled separately.

The characteristics of the trains defining the real behaviors and having a direct impact on the circulation are:

– the braking: the deceleration coefficient and the delay to activate the brake;

– the acceleration;

– the track speed profile: the maximum allowed speed depends on the train type.

With regard to the driver, the regulation theoretically guides the drivers but the behavior of the latter may sometimes be erratic. Consequently a typical behavior has been established by the experts. Here is an example, the “Vitesse Sécuritaire d’Approche (VISA)” (oncoming safety speed) principle in the BAL signaling: when a train approaches a warning signal, the driver must apply the VISA principle. This regulation requires the driver to brake as soon as he or she sees the signal in order to lower the speed to under 30 km/h, 200 m before the end of the block. The problem is then to know how the driver will perform this braking. Experience shows that in reality, the driver engages the braking when he or she sees the warning signal and so reaches the speed limit of 30 km/h at 300 m before the end of the block (instead of simply the end of block).

15.2.5. *Architecture of the model*

In addition to the description of two signaling systems, it is necessary to dynamically model the various operating modes of each safety instrumented systems concerned: the dynamic management of control systems for circulations protection (according to the requirements of the signaling systems), of the trains behavior (function of the equipments performances) and of the driver behavior. Given this complexity, a three level hierarchical architecture of the simulation model has been used. These levels are in the decreasing abstraction order as follows:

– Track: at this higher level, the track plan is represented. The problems related to the linking of successive blocks, the switch control and the itinerary interlocking are included at this level.

– Signaling: as the model must report the circulation of trains both with the ETCS system and with the BAL system and knowing that the driver behavior is different in these train types, it is important to know what signaling type is affected by the train in order to assign the adequate behavior. This selection is made at this level. Furthermore, as the lateral signaling is common to both systems, it will be managed at this level.

– Block: this level deals with the circulation, i.e. the speed management, the braking management, the management of the sojourn time of trains in a block, the calculus of the clearing time of the switches, etc.

As we already said, CPNs in CPN tools are well adapted to build hierarchical models due to the substitution transition concept. We then developed a generic and modular hierarchical modeling architecture based on submodels. Generic models for stations, blocks, switches, as well as signaling systems were developed. These models were then instantiated in order to get the whole model of a railway section of 40 km. A modular model whose granularity is representative of the physical elements of the system is thus obtained, allowing an easy readability in spite of its size.

15.2.6. *Example of an elementary model*

The modeled railway section, situated on the France-Luxembourg border and representative of the planed ERTMS corridors was chosen as a pilot project. It corresponds to approximately 40 km of tracks divided into 37 blocks and comprises 4 inputs, 3 outputs, 8 switches and 7 railway stations. The CPN tools model of this railway section contains about 550 places, 250 transitions, 20 different token colors and 70 functions written in (S)ML language.

As it is of course impossible to describe here this whole model, we will only present an example on timed CPN. Figure 15.13 gives this example, the management of the BAL signaling retransmission.

In CPNs, the tokens are differentiated; any token has a color containing information. A token modeling a railway block is characterized by a "block" color and will contain identifiers (block number and length), a token modeling a train will use a "train" color containing as identifiers the current speed and the maximum allowed speed (see Figure 15.14).

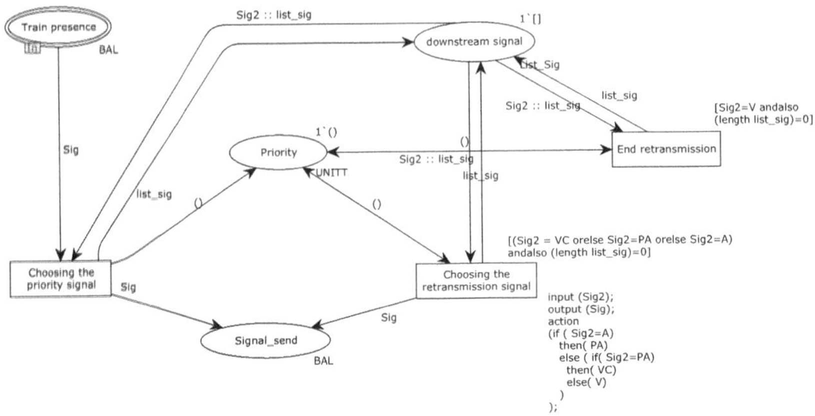

Figure 15.13. *Example of CTPN, transmission of the BAL signaling*

```
colset INT = int timed;
colset Train =record vmax : INT * v : INT * id : INT timed;
colset BAL =  with V| VC | PA | A ;
colset Canton = record longueur : INT * numero : INT;
colset List_Sig = list BAL ;
var train : Train;
var Sig, Sig2 : BAL;
var canton : Canton;
var list_sig : List_Sig ;
```

Figure 15.14. *Example of token statement*

A transition of the CPN will be characterized by several elements: a guard condition (Boolean expression that must be true in order to allow the firing of the transition), a time duration and a field to write ML code that will be executed on firing. Thus, it is possible to integrate in the model any algorithm to be taken into account for the study. For example, the calculus of the train speed and of its sojourn time in the block as a function of the entry speed, the state of signaling, the signaling type, the performances in acceleration and braking of the train, the declivity of the track, or the block length, etc.

15.2.7. *Incident generation*

The model being built of elementary bricks, it is easy to insert into it a perturbation at each abstraction level. It is possible to insert a signaling fault, a problem at the track level, mechanical failures of the train or a degradation of its performances or else a hazard in a driver behavior. These failures may all be associated with an occurrence probability. The injection of these perturbations in the model of the system gives access to the analysis of the robustness of the couple ETCS/BAL with regard to the failures of track and board components or even of people. As it is possible to know in real time the position of a train or its instantaneous speed, it is possible to predict how late it is regarding the schedule. After a simulation, campaign statistics can be established by the Monte-Carlo method.

We present here the results of the analysis obtained with two types of accident, each of them having a different back to normal phase (instantaneous or transitory):

– Track circuit out of order: track circuits are signaling elements in charge of the detection of train presence over a given area. Such a failure has the effect of signaling the block as being permanently occupied. Trains then have to go through the block in the running on sight mode. The driver must not exceed 30 km/h speed in order to be able to stop as soon as he observes an obstacle on the track.

– Breaking of the train coupling: this event leads to an immediate stop of the train which is then immobilized for the duration of the repair. This stop of course prevents the following trains continuing their progression. As soon as the repair is performed, the train starts again, to complete its itinerary and the trains, queued behind it, are restarted.

15.2.8. *Results*

Among the observable metrics, we chose to study the robustness of time schedules facing the occurrence of the aforementioned incidents (a schedule is the graphical representation of the train evolution in the space during the time). This schedule allows us to observe the ability of a signaling system to regain, or minimize, delays due to the occurrence of an feared event.

15.2.8.1. *Model validation*

Before using our model to obtain results that will assess the impact of the two signaling systems on the performance of the railway line, it is first necessary to ensure the proper design of the model, its good compliance relative to requirements and fidelity compared to the real behavior of the railway line. To achieve this, we chose to compare the schedules obtained by observation of the real line (data obtained from the Regional Centre of Observation of Nancy, corresponding to the circulation of a normal day with a nominal behavior, between 6 am and 9 am) and compare it to the schedules obtained by simulation of the developed CPN model.

Considering the BAL signaling system (the only one that is actually implemented in the real railway line), Figure 15.15 shows the real and simulated traffic schedules on the railway line between Zoufftgen and Woippy stations found on the considered pilot project, near to the France-Luxembourg border. The good superposition of the real and simulated curves shows the validity of our modeling approach.

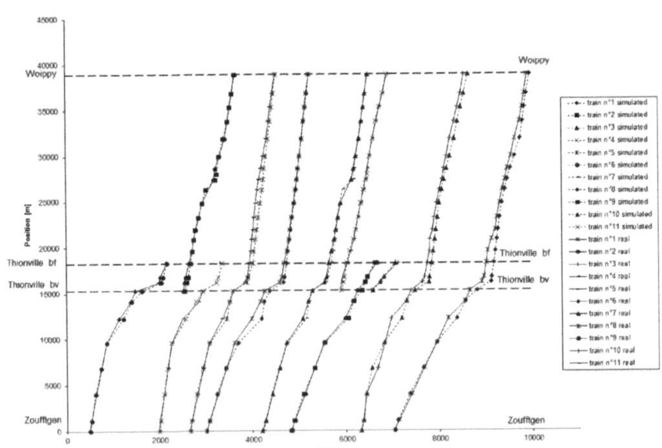

Figure 15.15. *Comparison of real and simulated schedules on the Zoufftgen-Woippy rail network*

We can now study the behavior of the line in downgraded modes. In downgraded behavior modes, we will consider 20 trains spaced every 3 min, which is the necessary time for the crossing of the largest block of the line at maximum speed. We then simulate this scenario once considering the ETCS signaling and once considering the BAL signaling.

15.2.8.2. *Failure of a track circuit*

The track circuit problem is inserted at kilometer 22. A lower drift is observed for the trains provided with the ETCS system (continuous line curve in Figure 15.16) than for trains provided with the BAL system (dotted line curve in Figure 15.16).

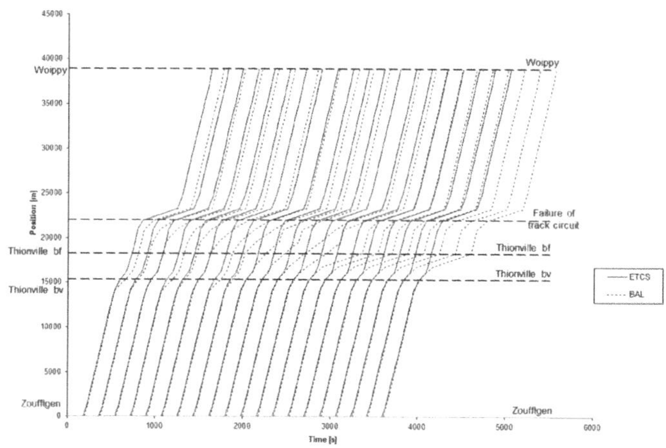

Figure 15.16. *Comparison ETCS/BAL in case of failure of a track circuit*

This can be explained by the braking phase optimization of the ETCS system in case of warning. Moreover, let us note that the BAL trains propagate the disturbance toward the previously crossed blocks unlike the ETCS trains that restrict the disturbance to the concerned area.

15.2.8.3. *Breaking of the train coupling*

Figure 15.17 shows the simulation results when a breaking of the train coupling occurs.

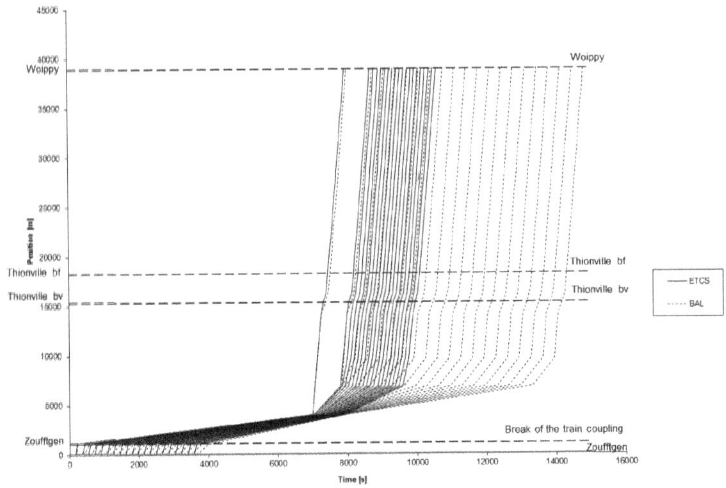

Figure 15.17. *Comparison ETCS/BAL in case of breaking of the train coupling*

Following this incident type (without transitory back to normal phase), one can clearly observe that ETCS trains allow a better recovery of the traffic towards a nominal situation. The recovery time to nominal situation for the BAL trains is more than two times greater than for ETCS trains.

15.2.8.4. *Other possible disturbances*

The architecture of the model makes it possible to get many other results without modification of the model itself. The impact of mixed traffic (alternation of BAL and ETCS trains) and the superposition of several failures have been especially evaluated.

15.2.9. *Conclusion*

The modeling power with CPN tools and particularly the modularity make possible the instantiation of any railway track by the simple modification of the parameters and insertion of additional elementary models. Any disturbance or failure may be inserted at any level in the model to evaluate their impact on the performances. As for the previous example, it is easy to bring such studies closer to the reliability concept. For such systems, the limits of the acceptable degradation of the delivered service correspond to the dysfunction of the system. Inserting the failures according to the probabilistic distribution of their occurrence time in a campaign of simulated stories, the Monte-Carlo method can give access to an evaluation of the reliability with regard to the considered failures. Of course, it is not a failure analysis method but only a powerful tool for the assessment of their impact. A preliminary study like FMEA must select the failure modes to be considered and potentially their superposition to evaluate the corresponding risk and to give an aid to the conception of prevention or protection barriers.

Conclusion

Petri nets (PNs) were not invented for the dependability study of systems. They first enjoyed success for the description and the design of discrete event systems until their declination integrating the stochastic aspect. Their equivalence with the continuous time Markov processes in case of exponential distribution of all the firing rates gave them a great interest for dependability studies because PNs allows us a progressive description of the phenomenon without requiring the *a priori* knowledge of all the states of the studied system. The emergence of advanced concepts in PNs increased their interest for the specialists in the dependability domain. The possibility of coexistence in a same model of immediate, delayed and stochastic transitions, substantially increased their interest. Notions of synchronization, interpretation, coloration brought significant means to model complex systems and solve difficult problems in the so-called dynamic reliability. We presented in this book a lot of these models and their application in several industrial fields to solve concrete problems. However, we never claimed to provide an exhaustive description neither of PN declinations nor of their applications in dependability domain. Many other and important contributions have extended the scope of PNs in dependability studies of complex systems. We will only cite the research of reachability of markings characterizing a feared state and of the set of possible corresponding scenarios or the problem of reverse reachability [MED 04, KHA 03, BOU 09]. We have also seen that specialists can use powerful tools including advanced concepts and allowing genericity, reusability, instanciability, etc. However, many safety and reliability analysts do not master the PN concepts. Perhaps this book will give them useful aid pending software developments for making their use transparent.

Appendix

A.1. Complements on Petri nets basics

A.1.1. *Other definitions*

– An unmarked Petri net (*PN*) *without structural conflict* is a PN in which any place has at most one output transition.

– A *structural conflict* in an unmarked PN will be denoted by the couple of the place and the transitions set concerned $< P_i, \{T_j, \ldots, T_m\} >$.

– A PN with *free choice* is such that for any conflict $< P_i, \{T_j, \ldots, T_m\} >$, none of the transitions $T_j, \ldots, T_m$ has another input place than P_i.

– A PN with *extended free choice* is such that for any conflict $< P_i, \{T_j, \ldots, T_m\} >$, all the transitions $T_j, \ldots, T_m$ have the same set of input places.

– A *PN simple* is such that each transition cannot be concerned by more than one conflict.

– An *effective conflict* is the existence of a structural conflict $< P_i, \{T_j, \ldots, T_m\} >$ and a marking M_c such that $M_c(P_i) < \sum w^+(P_i, T_k)$ where $T_k \in \{T_j, ..., T_k\}$ and T_k enabled by M_c.

– A PN is *effective conflict free* for a given initial marking M_0 if $\forall M_i \in R(M_0)$ there is no effective conflict.

– A PN is *persistent* for a given initial marking M_0 if $\forall M_i \in R(M_0)$ and if T_j and T_k are enabled by M_i, then T_j, T_k (or T_k, T_j) is a firing sequence from M_i.

– A PN is *pure* if it may be drawn from its incidence matrix (a PN in which the firing of a transition adds and retrieves tokens to a same place is not pure).

– A PN *without loops* is such that if a transition T_j and a place P_i exist and P_i is both input and output place of T_j , then T_j has at least another input place.

A.1.2. *Other properties*

– A PN is *safe* for a given initial marking M_0 if $\forall M_n \in R(M_0)$, $\forall P_i \in P$, $M_n(P_i) \leq 1$ (each place contains at most one token).

– A PN is *conform* if it is both live and safe.

– A transition is quasi-live for a given initial marking M_0 if: $\exists S$: $M_0 \xrightarrow{S} M_n / T_j \in S$ (there is at least one firing of T_j).

– A PN is *quasi-live* for a given initial marking M_0 if all its transitions are quasi-live for M_0.

– A PN has a *home state* M_a for a given initial marking M_0 if: $\forall M_n \in R(M_0), \exists S : Mn \xrightarrow{S} M_a$.

This state is dependent on the initial marking and for a given initial marking, many home places can exist; they constitute the home space for M_0.

– A PN is *resettable* for M_0 if M_0 is a home state.

A.1.3. *Special configurations*

A.1.3.1. *State graph and event graph*

– An unmarked PN is a *state graph* (or state automaton) if and only if any transition has exactly a single input place and a single output place (Figure A.1). When marked, such a PN has an equivalent behavior as a state graph if and only if it contains a single token.

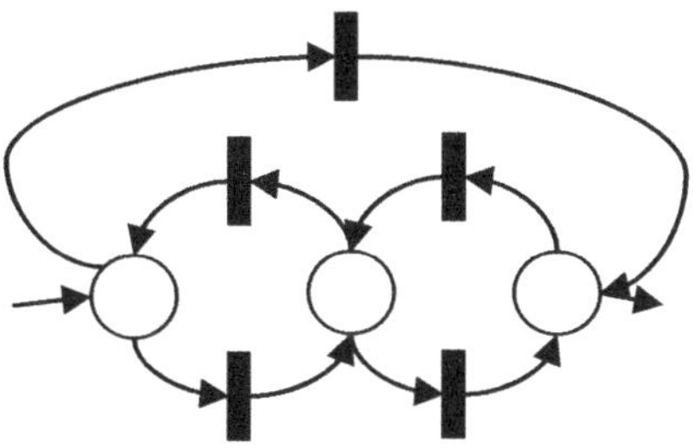

Figure A.1. *State graph*

– A PN is an *event graph* when each place has exactly a single input transition and a single output transition (Figure A.2). It is the dual of a state graph.

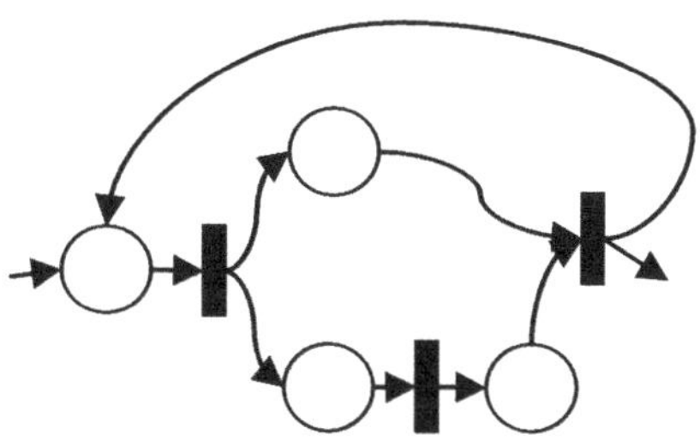

Figure A.2. *Event graph*

A.1.3.2. *Lock and trap*

Let us consider a subset of places in an ordinary PN: $P\prime = \{P1, P2, P3, \ldots, Pr\} \subseteq P$ and the notation:

$I(P\prime) = I(P_1) \cup I(P_2) \cup I(P_3) \cup \ldots \cup I(P_r)$ and $O(P\prime) = O(P_1) \cup O(P_2) \cup O(P_3) \cup \ldots \cup O(P_r)$.

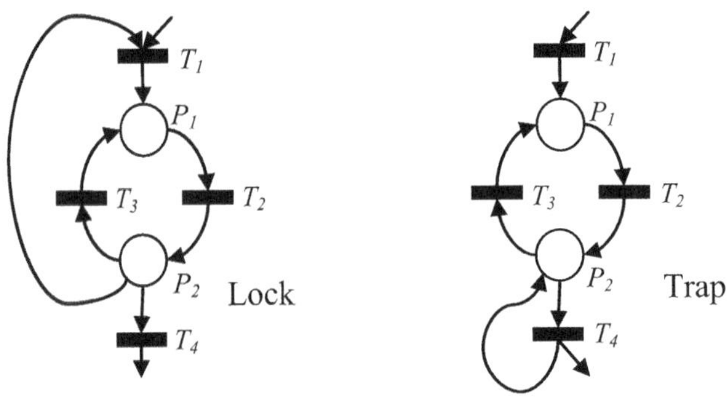

Figure A.3. *Lock and trap in a PN*

– A lock (Figure A.3(left)) is a subset of places so that the set of their input transitions is included in the set of their output transitions: $I(P\prime) \subseteq O(P\prime)$

– A trap (Figure A.3(right)) is a subset of places so that the set of their output transitions is included in the set of their input transitions: $O(P\prime) \subseteq I(P\prime)$

Bibliography

[AIS 15] AIS GROUP EINDHOVEN UNIVERSITY OF TECHNOLOGY T.N., CPN Tools software, available at http://cpntools.org/, 2015.

[AJM 84] AJMONE-MARSAN M., BALBO G., CONTE G., "A class of generalized stochastic Petri nets for the performance evaluation of multiprocessor systems", *ACM Transactions on Computer Systems*, vol. 2, no. 2, pp. 93–122, May 1984.

[AJM 86] AJMONE-MARSAN M., CHIOLA G., "On Petri nets with deterministic and exponentially distributed firing times", *Lectures Notes in Computer Science*, vol. 266, pp. 132–145, 1986.

[AJM 89] AJMONE-MARSAN M., BALBO G., BOBIO A. *et al.*, "The effect of execution policies on the semantics and analysis of stochastic Petri nets", *IEEE Transactions on Software Engineering*, IEEE, vol. 15, no. 7, pp. 832–846, 1989.

[AQS 96] AQS-GT OORS, (Association Qualité-Sécurité (AQS) – Groupe de travail Observatoire de l'Opinion sur les Risques de la Sécurité), Management de la sécurité d'entreprise, vocabulaire et concepts, March 1996.

[ATW 80] ATWOOD C.L., Estimating common cause failure rates for pumps in nuclear reactors, PhD Thesis, California, 1980.

[AUB 87] AUBRY J.-F., Conception des systèmes de commande numériques des convertisseurs électromécaniques: vers une méthodologie intégrant la Sûreté de fonctionnement, PhD Thesis, Institut National Polytechnique de Lorraine, 1987.

[AUB 91] AUBRY J.-F., ZANNE C., "Intégration de la sûreté de fonctionnement dans la conception des systèmes de commande des processus électromécaniques", *Automatique Productique Informatique Industrielle APII*, pp. 297–324, 1991.

[AUB 08] AUBRY J.-F., CHATELET E., "Sûreté de fonctionnement des systèmes de commande: exemple d'application et rappels sur les réseaux de Petri", *Les techniques de l'ingénieur. Traité Informatique Industrielle*, systèmes d'information et de communication, no. S8263, p. 6, 2008.

[AUB 12] AUBRY J.-F., BABYKINA G., BARROS A. *et al.*, Rapport final du projet APPRODYN: "APPROches de la fiabilité DYNamique pour modéliser des systèmes critiques", Report, collaboration CRAN, EDF R&D, INRIA-CQFD, UTT-ICD. available at http://hal.archives-ouvertes.fr/hal- 00740181/PDF/ Rapport_final_APPRODYN_v7a_NB.pdf, 2012.

[AUB 15] AUBRY J.-F., BRINZEI N., *Systems Dependability Assessment: Modeling with Graphs and Finite State Automata*, ISTE, London and John Wiley & Sons New York, April 2015.

[BAB 03] BABAK A.-S., "Stability of networked control systems in the presence of packet losses", *IEEE Conference on Decision and Control*, Maui, Hawaii, pp. 676–681, December 2003.

[BAR 09] BARGER P., SCHON W., BOUALI M., "A study of railway ERTMS safety with colored Petri nets", *Annual Conference of the European Safety and Reliability Association, ESREL 2009*, Praha, vol. 2, pp. 1303–1309, September 2009.

[BES 01] BEST E., DEVILLERS R., KOUTNY M., *Petri Net Algebra*, Springer, 2001.

[BOB 86] BOBBIO A., TRIVEDI K., "An aggregation technique for the transient analysis of stiff Markov chains", *IEEE Transactions on Computers*, IEEE, vol. C-35, no. 2, pp. 803–814, September 1986.

[BOU 09] BOUALI M., BARGER P., SCHON W., "Colored Petri nets inversion for backward reachability analysis", *Second IFAC Workshop on Dependable Control of Discrete Systems*, Bari, Italy, IFAC, pp. 259–264, June 2009.

[BRA 83] BRAMS G.-W., *Réseaux de Petri Théorie et pratique*, Masson, 1983.

[BRA 03] BRANICKY M.-S., LIBERATORE V., PHILIPS S.-M., "Networked control system co-simulation for co-design", *Proceedings of the American Control Conference. ACC 2003,* Denver, Colorado, American Automatic Control Council, pp. 3341–3346, June 2003.

[BRI 13] BRICMAN-REJC Z., CEPIN M., SITDIKOVA A., "Estimating common-cause failures parameters within power system reliability analysis", *Annual Conference of the European Safety and Reliability Association, ESREL 2013*, Amsterdam, pp. 2841–2846, 2013.

[BRI 14] BRINZEI N., DELEUZE G., VILLAUME N. *et al.*, "Common cause failures modelling by means of coloured Petri nets for dependability assessment of a control system of nuclear power plant", *Safety and Reliability: Methodology and Applications. Proceedings of European Safety and Reliability Conference ESREL 2014*, CRC Press/Balkema, Taylor & Francis Group, pp. 2121–2129, September 2014.

[CAS 99] CASSANDRAS C., LAFORTUNE S., *Introduction to Discrete Event Systems,* 1st ed., Kluwer Academic Publishers, 1999.

[CAS 08] CASSANDRAS C., LAFORTUNE S., *Introduction to Discrete Event Systems,* 2nd ed., Springer Verlag, 2008.

[CHA 98] CHABOT J.-L., Approche probabiliste relative à l'étude des scénarios d'incendie, PhD Thesis, Université de Poitiers France, October 1998.

[CHA 99] CHABOT J.-L., DUCAMP F., DUTUIT Y. *et al.*, "Simulation hybride, méthode de modélisation intégrant phénomènes continus et discrets", *Proceedings of 11ème Colloque National de Fiabilité et maintenabilité - $\lambda\mu$ 11* - Arcachon, France, pp. 126–136, 1999.

[CHA 00] CHARLET J., ZACKLAD M., KASSEL G. *et al.*, *Ingénierie des connaissances, évolutions récentes et nouveaux défis*, Eyrolles, 2000.

[CHI 93] CHIOLA G., AJMONE-MARSAN M., BALBO G. *et al.*, "Generalized stochastic Petri nets: a definition at the net level and its implications", *IEEE Transactions on Software Engineering*, IEEE, vol. 19, no. 2, pp. 89–107,1993.

[CHO 93a] CHOI H., Performance and reliability modeling using Markov regenerative stochastic Petri nets, PhD Thesis, Duke University, Durham, 1993.

[CHO 93b] CHOI H., KULKARNI V., TRIVEDI K.-S., "Transient analysis of deterministic and stochastic Petri nets", *Lecture Notes in Computer Science. Proceedings of 14th International Conference on Application and Theory of Petri Nets*, Springer-Verlag, no. 691, pp. 166–185, 1993.

[CHO 94] CHOI H., KULKARNI V., TRIVEDI K.-S., "Markov regenerative stochastic Petri nets", *Performance Evaluation*, vol. 20, nos. 1–3, pp. 337–357, 1994.

[CHR 83] CHRÉTIENNE P., Les réseaux de Petri temporisés, PhD Thesis, University of Paris VI, June 1983.

[DAN 95] DANIEL O., Les réseaux de Petri stochastiques pour l'évaluation des attributs de la sûreté de fonctionnement des systèmes manufacturiers, PhD Thesis, Institut National Polytechnique de Grenoble, 1995.

[DAV 89] DAVID R., ALLA H., *Du Grafcet aux réseaux de Petri*, Hermes, 1989.

[DAV 92] DAVID R., ALLA H., *Petri Nets and Grafcet. Tools for Modeling Discrete Event Systems*, Prentice Hall, 1992.

[DEL 15a] DELEUZE G., BRINZEI N., "Modeling digital I&C systems for PRA with coloured Petri nets", *9th International Conference on Nuclear Plant Instrumentation, Control & Human-Machine Interface Technologies, NPIC&HMIT 2015*, February 2015.

[DEL 15b] DELEUZE G., BRINZEI N., GÉRARD L., "Common cause failures parameters estimation with coloured Petri nets", *International Topical Meeting on Probabilistic Safety Assessment and Analysis PSA 2015*, CdRom paper number 11611, April 2015.

[DIA 01] DIAZ M., *Les réseaux de Petri, modèles fondamentaux*, Hermes, 2001.

[DUG 84] DUGAN J., TRIVEDI K.-S., GEIST R.-M. *et al.*, "Extended stochastic Petri nets: applications and analysis", *10th International Symposium on Computer Performance,* Paris, pp. 507–519, 1984.

[DUT 97] DUTUIT Y., CHATELET E., SIGNORET J. *et al.*, "Dependability modelling and evaluation by using stochastic Petri nets: application to two test cases", *Reliability Engineering & System Safety*, Elsevier, vol. 55, no. 2, pp. 117–124, 1997.

[ERT 99] EEIG ERTMS, User group ERTMS/ETCS system requirements specification, UIC (International Union of Railways) Brussels, 1999.

[GHO 08a] GHOSTINE R., Influence des fautes transitoires sur la fiabilité d'un système commandé en réseau, PhD Thesis, Institut National Polytechnique de Lorraine, June 2008.

[GHO 08b] GHOSTINE R., THIRIET J., AUBRY J., "Framework for reliability evaluation of networked control systems", *17th IFAC World Congress*, Seoul, Korea, IFAC, p. 6, July 2008.

[GHO 11] GHOSTINE R., THIRIET J., AUBRY J., "Variable delays and message losses: influence on the reliability of a control loop", *Reliability Engineering & System Safety*, Elsevier, vol. 96, no. 1, pp. 160–171, January 2011.

[GRU 93] GRUBER T.-R., THOMAS R., *Towards Principles for the Design of Ontologies Used for Knowledge Sharing in Formal Ontology in Conceptual Analysis and Knowledge Representation*, Kluwer Academic Publishers, 1993.

[GRU 15] GRUNT O., BRIS R., "SPN as a tool for risk modeling of fires in process industries", *Journal of Loss Prevention in the Process Industries*, Elsevier, vol. 34, pp. 72–81, 2015.

[GT 03] Groupe de Travail Méthodologie, Principes généraux pour l'élaboration et la lecture des études de dangers, INERIS Intitut National de l'Environnement Industriel et des Risques, 2003.

[GUA 98] GUARINO N., *Some Ontological Principles for Designing Upper Level Lexical Resources*, IOS Press, 1998.

[HAR 00] HARPER R., *Programming in standard ML*, Carnegie Mellon University, 2000.

[HOR 98] MEYER ZU HORSTE M., SCHNIEDER E., "Modelling and simulation of train control systems using Petri nets, FM 99 formal methods", *FM 99 Formal Methods. World Congress on Formal Methods in the Development of Computing Systems.*, vol. 1709, pp. 1867–1883, 1998.

[HSE 92] HSE, Generic terms and concepts in the assessment and regulation of industrial risks, Health and Society Executive, 1992.

[IEC 10] IEC-61508, Functional safety of electrical/electronic/ programmable electronic safety-related, vol. 1–7, 2010.

[IEC 11] IEC-61513, Nuclear power plants – instrumentation and control for systems important to safety – general requirements for systems, 2011.

[IEC 90] IEC 60050-191 International electrotechnical vocabulary chapter 191: dependability and quality of service, December 1990.

[IEC 02] IEC 62278 International Standard – railway applications – specification and demonstration of reliability, availability, maintainability and safety (RAMS), September 2002.

[IEC 07] IEC 62425 International Standard – railway applications – communication, signalling and processing systems – safety related electronic systems for signalling, September 2007.

[IEC O2] IEC 62279 International Standard – railway applications – communication, signalling and processing systems – software for railway control and protection systems, September 2002.

[IEE 04] IEEE standard for communications-based train control (CBTC) performance and functional requirements, 2004.

[ISO 02] ISO, Management du risque Vocabulaire principes directeurs pour les inclure dans les normes, 2002.

[ISO 14] ISO 21127:2014 – information and documentation – a reference ontology for the interchange of cultural heritage information, 2014.

[JAM 00] JAMPI D., AUBRY J.-F., PORRAS J. *et al.*, "Spécification de systèmes mécatroniques sûrs de fonctionnement", *Proceedings of 12ème Colloque National de Fiabilité et maintenabilité – λμ 11*, pp. 521–528, Montpellier, France, 2000.

[JAM 01] JAMPI D., Détermination d'une méthodologie d'aide à la conception d'un système de contrôle commande numérique sûr de fonctionnement, PhD Thesis, Institut National Polytechnique de Lorraine, October 2001.

[JAN 98] JANSEN L., MEYER ZU HÖRSTE M., SCHNIEDER E., "Technical issues in modeling the European train control system (ETCS) using coloured Petri nets and the Design/CPN tools", *Proceedings of the Workshop on Practical Use of Coloured Petri Nets and Design/CPN*, Aarhus, Denmark, pp. 103–115, 1998.

[JEN 95] JENSEN K., *Coloured Petri Nets: Basic Concepts, Analysis Methods and Practical Use. Vol. 2, Analysis Methods*, Springer-Verlag, 1995.

[JEN 97a] JENSEN K., *Coloured Petri Nets: Basic Concepts, Analysis Methods and Practical Use. Vol 1, Basic Concepts. 2nd ed.*, Springer-Verlag, 1997.

[JEN 97b] JENSEN K., *Coloured Petri Nets: Basic Concepts, Analysis Methods and Practical Use. Vol. 3, Practical Use*, Springer-Verlag, 1997.

[JEN 09] JENSEN K., KRISTENSEN L., *Coloured Petri Nets: Modelling and Validation of Concurrent Systems*, Springer-Verlag, 2009.

[KER 00] KERMISCH C., LABEAU P.-E., Approche dynamique de la fiabilité des systèmes, Rapport de projet ISdF, Report, 2000.

[KHA 03] KHALFAOUI S., Méthode de recherche des scénarios redoutés pour l'évaluation de la sûreté de fonctionnement des systèmes mécatroniques du monde automobile, PhD Thesis, Institut National Polytechnique de Toulouse, September 2003.

[KOK 99] KOKOTOVIC P., KHALIL H.-K., O'REILLY J., *Singular Perturbation Methods in Control, Analysis and Design*, SIAM, 1999.

[LAH 06] LAHLOU O., EL-KOURSI E., BON P. *et al.*, "Evaluation des règles d'exploitation pour l'interopérabilité et la sécurité dans les transports ferroviaires", *6ème Conférence Francophone de Modéisation et Simulation, MOSIM 06*, vol. 1, pp. 291–297, Rabat, Morocco, 2006.

[LAL 10a] LALOUETTE G., BRINZEI N., MALASSÉ O. *et al.*, "Evaluation des performances du système de signalisation ferroviaire européen superposé au système français en présence de défaillances", *Congrès Lambda Mu 17 Maîtrise des Risques et Sûreté de Fonctionnement,* La Rochelle, France, CdRom paper number 5D-1, October 2010.

[LAL 10b] LALOUETTE G., BRINZEI N., MALASSÉ O. *et al.*, "Modélisation et évaluation des performances d'un système de signalisation ferroviaire intégrant BAL et ETCS par réseaux de Petri colorés", *Sixième Conférence Internationale Francophone d'Automatique, CIFA 2010 – France*, CdRom paper number 111, June 2010.

[LIN 98] LINDEMANN C., *Performance Modelling with Deterministic and Stochastic Petri Nets*, John Wiley & Sons, 1998.

[MAR 96] MARSEGUERRA M., ZIO E., "Monte Carlo approach to PSA for dynamic process systems", *Reliability Engineering & System Safety*, Elsevier, vol. 52, no. 3, pp. 227–241, 1996.

[MAZ 07] MAZOUNI M.-H., BIED-CHARRETON D., AUBRY J.-F., "Proposal of a generic methodology to harmonize preliminary hazard analyses for guided transport", *Proceedings of IEEE SMC International Conference SOSE'2007: System of Systems Engineering*, IEEE, San Antonio, Texas , no. 98, April 2007.

[MAZ 08] MAZOUNI M.-H., Pour une Meilleure Approche du Management des Risques: De la Modélisation Ontologique du Processus Accidentel au Système Interactif d'Aide Ã la Décision, PhD Thesis, Institut National Polytechnique de Lorraine, November 2008.

[MED 04] MEDJOUDJ M., KHALFAOUI S., DEMMOU H. *et al.*, "A method for deriving feared scenarios in hybrid systems", *Probabilistic Safety Assessment and Management (PSAM'7 – ESREL'04)*, Berlin, Germany, pp. 1870–1875, 2004.

[MOA 78a] MOALLA M., PULOU J., SIFAKIS J., "Réseaux de Petri synchronisés", *RAIRO Automatique/Systems Analysis and Control*, EDP Sciences, vol. 12, no. 2, pp. 103–130, 1978.

[MOA 78b] MOALLA M., PULOU J., SIFAKIS J., "Synchronized petri nets: a model for the description of non-autonomous systems", *Mathematical Foundations of Computer Science*, Springer, Berlin, Heidelberg, vol. 58, pp. 374–384, 1978.

[MOA 80] MOALLA M., SIFAKIS J., SILVA M., A la recherche d'une méthodologie de conception sûre des automatismes logiques basée sur l'utilisation des réseaux de Petri, Monographie AFCET: Sureté de Fonctionnement des Systèmes Informatiques, 1980.

[MOA 85] MOALLA M., "Réseaux de Petri interprétés et Grafcet", *Technique et Science Informatique*, vol. 4, no. 1, pp. 103–130, 1985.

[MOI 91] MOITESSIER F., MISTRAL: Méthodologie Interactive de conception et réalisation des Systèmes de commande Temps Réel répartis en Automatique: application à un processus électromécanique, PhD Thesis, Institut National Polytechnique de Lorraine, 1991.

[PER 07] PÉRILHON P., *La Gestion des risques: Méthode MADS-MOSAR II*, Editions Demos, Paris, 2007.

[PET 62] PETRI C.-A., Kommunikation mit Automaten, PhD Thesis dissertation, Rheinisch-Westfälisches Institut fur instrumentelle Mathematik an der Universitat Bonn., Nr. 2, 1962.

[POZ 04a] POZSGAI P., BERTSCHE B., "Conjoint modelling with extended coloured stochastic Petri net and reliability block diagram for system analysis", *Probabilistic Safety Assessment and Management PSAM 7/European Safety and Reliability Conference ESREL'04*, Berlin, Germany, Springer, pp. 1382–1387, 2004.

[POZ 04b] POZSGAI P., BERTSCHE B., "Integrating the age development and operational costs into a comprehensive system analysis modellingprocedure", *Lifetime Management of Industrial Systems: Proceedings of the 26th ESReDA Seminar,* European Commission Publisher, Tampere, Finland, 2004.

[RAM 74] RAMCHANDANI C., Analysis of asynchronous concurrent systems by timed Petri nets, PhD Thesis, MIT Cambridge Mass.Project MAC MAC-TR120., 1974.

[RFF 07] RFF, Corridors européens ETCS1. Cahier de charge fonctionnel. Principes de signalisation ETCS1, 2007.

[SÉM 97] SÉMANTIQUES DU RISQUE G.A., Vocabulaire lié au risque à travers une analyse bibliographique, 1997.

[ROC 05] ROCHE C., "Ontologie et Terminologie", *Larousse revue*, Larousse, no. 157, pp. 1–11, 2005.

[SAT 15] SATODEV, Grif-workshop: graphical interface for reliability forecasting, available at http://grif-workshop.com/, 2015.

[SCH 03] SCHOENIG R., AUBRY J.-F., CAMBOIS T. *et al.*, "A design methodology for embedded control systems including safety assessment studies", *ESREL European Safety and RELiabilty Conference,* Maastricht, The Netherlands, pp. 1401–1408, June 2003.

[SCH 04] SCHOENIG R., Définition d'une méthodologie de conception des systèmes mécatroniques sûrs de fonctionnement, PhD Thesis, Institut National Polytechnique de Lorraine, France, October 2004.

[SCH 06] SCHOENIG R., AUBRY J.-F., CAMBOIS T. *et al.* , "An aggregation method of Markov graphs for the reliability analysis of hybrid systems", *Reliability Engineering and System Safety*, Elsevier, vol. 91, no. 2, pp. 137–148, February 2006.

[SEV 96] DIRECTIVE 96/82/EC (SEVESO II) On the control of major-accident hazards involving dangerous substances, December 1996.

[SIF 77] SIFAKIS J., "Use of Petri nets for performance evaluation", *3rd International Symposium on Modeling and Evaluation, IFIP*, North Holland, pp. 75–93, 1977.

[SIF 80] SIFAKIS J., "Performance evaluation of systems using Petri nets", *Lecture Notes in Computer Science*, vol. 84, pp. 307–319, 1980.

[SOG 06] SOGLO A., XIANHUI Y., "Networked control system simulation design and its application", *Tsinghua Science And Technology*, Elsevier, vol. 11, no. 3, pp. 287–294, June 2006.

[TAN 88] TANKOANO J., M2C: Une approche méthodique pour la conception certifiée des systèmes de commande des automatismes industriels répartis, PhD Thesis, Université de Nancy 1, 1988.

[TOR 08] DI TORINO U., GRaphical editor and analyzer for timed and stochastic Petri nets, available at http://www.di.unito.it/~greatspn/index.html, 2008.

[USN 87] US-NRC, NUREG/CR-4780. Procedures for treating common-cause failures in safety and reliability studies, US Nuclear Regulatory Commission, vol. 1 and 2, Washington, DC, 1987.

[ZEI 14] ZEILER P., BERTSCHE B., "Availability modelling and analysis of an offshore wind turbine using extended coloured stochastic Petri nets", *Safety and Reliability: Methodology and Applications. Proceedings of European Safety and Reliability Conference ESREL 2014*, CRC Press/Balkema, Taylor & Francis Group, pp. 1157–1165, September 2014.

[ZHA 01] ZHANG W., BRANICKY M., PHILIPS S., "Stability of networked control systems", *IEEE Control Systems Magazine*, IEEE, vol. 20, no. 1, pp. 84–99, 2001.

[ZHA 08] ZHANG H., GONZALEZ K., DUFOUR F. *et al.*, "Piecewise deterministic Markov processes and dynamic reliability", *Proceedings of the Institution of Mechanical Engineers, Part O: Journal of Risk and Reliability*, vol. 222, no. 4, pp. 545–551, 2008.

[ZIM 03] ZIMMERMANN A., HOMMEL G., "A train control system case study in model-based real time system design", *In proc. 11th Intern. Workshop on Parallel & Distributed Real-Time Systems, (in conjunction with IEEE – IPDPS 03)*, Nice, France, p. 8, 2003.

Index

M, N, O

P, R, S

T, U, V

Other titles from

iSTE

in

Systems and Industrial Engineering

2015

AUBRY Jean-François, BRINZEI Nicolae
Systems Dependability Assessment: Modeling with Graphs and Finite State Automata

BOULANGER Jean-Louis
CENELEC 50128 and IEC 62279 Standards

BRIFFAUT Jean-Pierre
E-Enabled Operations Management

MISSIKOFF Michele, CANDUCCI Massimo, MAIDEN Neil
Enterprise Innovation

2014

CHETTO Maryline
Real-time Systems Scheduling
Volume 1 – Fundamentals
Volume 2 – Focuses

DAVIM J. Paulo
Machinability of Advanced Materials

ESTAMPE Dominique
Supply Chain Performance and Evaluation Models

FAVRE Bernard
Introduction to Sustainable Transports

MICOUIN Patrice
Model Based Systems Engineering: Fundamentals and Methods

MILLOT Patrick
Designing Human–Machine Cooperation Systems

NI Zhenjiang, PACORET Céline, BENOSMAN Ryad, REGNIER Stéphane
Haptic Feedback Teleoperation of Optical Tweezers

OUSTALOUP Alain
Diversity and Non-integer Differentiation for System Dynamics

REZG Nidhal, DELLAGI Sofien, KHATAD Abdelhakim
Joint Optimization of Maintenance and Production Policies

STEFANOIU Dan, BORNE Pierre, POPESCU Dumitru, FILIP Florin Gh.,
EL KAMEL Abdelkader
Optimization in Engineering Sciences: Metaheuristics, Stochastic Methods andDecision Support

2013

ALAZARD Daniel
Reverse Engineering in Control Design

ARIOUI Hichem, NEHAOUA Lamri
Driving Simulation

CHADLI Mohammed, COPPIER Hervé
Command-control for Real-time Systems

DAAFOUZ Jamal, TARBOURIECH Sophie, SIGALOTTI Mario
Hybrid Systems with Constraints

FEYEL Philippe
Loop-shaping Robust Control

FLAUS Jean-Marie
Risk Analysis: Socio-technical and Industrial Systems

FRIBOURG Laurent, SOULAT Romain
Control of Switching Systems by Invariance Analysis: Application to Power Electronics

GRUNN Emmanuel, PHAM Anh Tuan
Modeling of Complex Systems: Application to Aeronautical Dynamics

HABIB Maki K., DAVIM J. Paulo
Interdisciplinary Mechatronics: Engineering Science and Research Development

HAMMADI Slim, KSOURI Mekki
Multimodal Transport Systems

JARBOUI Bassem, SIARRY Patrick, TEGHEM Jacques
Metaheuristics for Production Scheduling

KIRILLOV Oleg N., PELINOVSKY Dmitry E.
Nonlinear Physical Systems

LE Vu Tuan Hieu, STOICA Cristina, ALAMO Teodoro,
CAMACHO Eduardo F., DUMUR Didier
Zonotopes: From Guaranteed State-estimation to Control

MACHADO Carolina, DAVIM J. Paulo
Management and Engineering Innovation

MORANA Joëlle
Sustainable Supply Chain Management

SANDOU Guillaume
Metaheuristic Optimization for the Design of Automatic Control Laws

STOICAN Florin, OLARU Sorin
Set-theoretic Fault Detection in Multisensor Systems

2012

AÏT-KADI Daoud, CHOUINARD Marc, MARCOTTE Suzanne, RIOPEL Diane
Sustainable Reverse Logistics Network: Engineering and Management

BORNE Pierre, POPESCU Dumitru, FILIP Florin G., STEFANOIU Dan
Optimization in Engineering Sciences: Exact Methods

CHADLI Mohammed, BORNE Pierre
Multiple Models Approach in Automation: Takagi-Sugeno Fuzzy Systems

DAVIM J.Paulo
Lasers in Manufacturing

DECLERCK Philippe
Discrete Event Systems in Dioid Algebra and Conventional Algebra

DOUMIATI Moustapha, CHARARA Ali, VICTORINO Alessandro, LECHNER Daniel
Vehicle Dynamics Estimation using Kalman Filtering: Experimental Validation

HAMMADI Slim, KSOURI Mekki
Advanced Mobility and Transport Engineering

MAILLARD Pierre
Competitive Quality Strategies

MATTA Nada, VANDENBOOMGAERDE Yves, ARLAT Jean
Supervision and Safety of Complex Systems

POLER Raul *et al.*
Intelligent Non-hierarchical Manufacturing Networks

YALAOUI Alice, CHEHADE Hicham, YALAOUI Farouk, AMODEO Lionel
Optimization of Logistics

ZELM Martin *et al.*
Enterprise Interoperability –I-EASA12 Proceedings

2011

CANTOT Pascal, LUZEAUX Dominique
Simulation and Modeling of Systems of Systems

DAVIM J. Paulo
Mechatronics

DAVIM J. Paulo
Wood Machining

KOLSKI Christophe
Human–Computer Interactions in Transport

LUZEAUX Dominique, RUAULT Jean-René, WIPPLER Jean-Luc
Complex Systems and Systems of Systems Engineering

ZELM Martin, *et al.*
Enterprise Interoperability: IWEI2011 Proceedings

2010

BOTTA-GENOULAZ Valérie, CAMPAGNE Jean-Pierre, LLERENA Daniel, PELLEGRIN Claude
Supply Chain Performance / Collaboration, Alignement and Coordination

BOURLÈS Henri, GODFREY K.C. Kwan
Linear Systems

BOURRIÈRES Jean-Paul
Proceedings of CEISIE'09

DAVIM J. Paulo
Sustainable Manufacturing

GIORDANO Max, MATHIEU Luc, VILLENEUVE François
Product Life-Cycle Management / Geometric Variations

LUZEAUX Dominique, RUAULT Jean-René
Systems of Systems

VILLENEUVE François, MATHIEU Luc
Geometric Tolerancing of Products

2009

DIAZ Michel
Petri Nets / Fundamental Models, Verification and Applications

OZEL Tugrul, DAVIM J. Paulo
Intelligent Machining

2008

ARTIGUES Christian, DEMASSEY Sophie, NÉRON Emmanuel
Resources–Constrained Project Scheduling

BILLAUT Jean-Charles, MOUKRIM Aziz, SANLAVILLE Eric
Flexibility and Robustness in Scheduling

DOCHAIN Denis
Bioprocess Control

LOPEZ Pierre, ROUBELLAT François
Production Scheduling

THIERRY Caroline, THOMAS André, BEL Gérard
Supply Chain Simulation and Management

2007

DE LARMINAT Philippe
Analysis and Control of Linear Systems

LAMNABHI Françoise *et al.*
Taming Heterogeneity and Complexity of Embedded Control

LIMNIOS Nikolaos
Fault Trees

2006

NAJIM Kaddour
Control of Continuous Linear Systems

Printed and bound by CPI Group (UK) Ltd, Croydon, CR0 4YY
13/09/2026
14961055-0002